DIY
CENTRAL
HEATING

DIY CENTRAL HEATING

Written and illustrated by
ARTHUR BAKER

Devised by
Alan Rice

Edited by
David Thomas

ORBIS · LONDON

Contents

6 **INTRODUCTION**

9 **HOW CENTRAL HEATING WORKS**
10 Fully automatic gas central heating
12 Gas boilers
14 Solid-fuel boilers
15 Oil boilers
16 Solid-fuel room-heaters
18 Gas room-heaters
19 Wood-fired boilers
20 Cooker/boilers
22 Panel radiators
23 Convector and skirting radiators
24 Capillary fittings
26 Compression fittings
28 Pumps
30 Central-heating controls
32 Microbore
33 Cylinder conversion unit

35 **DESIGNING A SYSTEM**
36 Central-heating design
40 Sizing radiators
41 Drawing plans
46 Making an isometric drawing
47 Sizing the pipes

49 The index circuit
52 Solid-fuel system
54 Designing for microbore
58 Understanding wiring diagrams
60 Designing for pumps and control systems
62 Basic design layouts

65 **INSTALLING CENTRAL HEATING**
66 Tools to buy
67 Tools to hire
67 Bending pipe
68 Fitting radiators
70 Joining pipe with capillary fittings
71 Joining pipe with compression fittings
71 Threaded fittings
72 Installing the boiler/solid-fuel room-heater
74 Installing the boiler/wall-hung gas boiler
76 Installing a flue-liner
77 Oil storage tank installation
78 Removing floorboards
79 Running pipes beneath floors
80 Running the pipes
82 The hot-water cylinder

84 The feed and expansion tank
86 Fitting the pump
88 Fitting control valves
90 The wiring
92 Microbore installation
94 Order of work and system checklist
95 Finishing off
96 Radiators for difficult areas
97 Fitting a new boiler
97 Decorating behind radiators
98 Faults in central-heating systems
98 Extensions
98 Replacing a radiator
99 Replacing a pump
100 Replacing a cold-water storage tank
101 Insulation
102 Chimneys and flues
103 The regulations
104 Heat-loss calculation charts
106 Itemized checklist
108 Scrap disposal
108 Keeping records
109 Glossary
111 Acknowledgments
112 Index

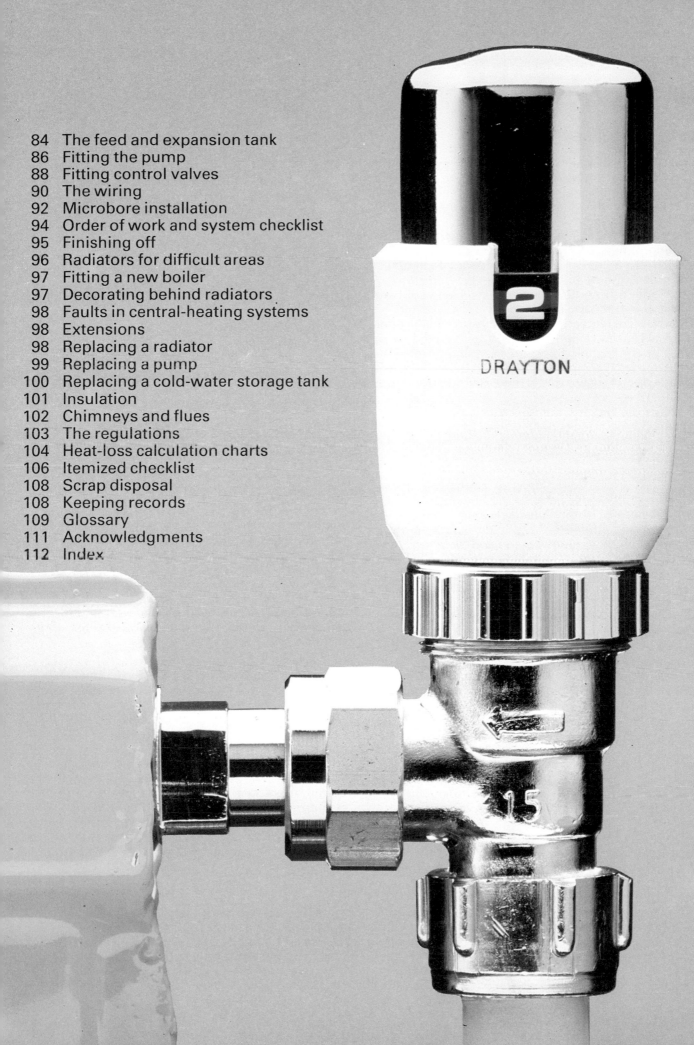

Introduction

DOING IT YOURSELF

Can you do it yourself? Installing your own central heating is not a weekend job but there are a lot more ambitious projects attempted by the do-it-yourselfer and it is well within the capabilities of a careful and competent person.

There is a satisfaction about working with modern plumbing fittings and making neat runs of pipe. The materials are all new and tailor-made for each other, and the pleasure of an efficiently heated home, many winters after having designed and installed your own system, is immense.

The main advantage of doing it yourself is economic: you can halve the price you are quoted for a professionally installed system by doing it yourself. Further advantages are that the home-owner can take greater care of decorations and disrupt the room furnishings as little as possible.

Perhaps the biggest difference between the home installer and the professional is time. Of necessity, the contractor is working against the clock, whereas the home-owner can take as much time as he likes, within reason, to get a perfect result.

But there are disadvantages of course. It's heavy work. There are some parts of the installation which require muscle – locating precisely a wall-hung boiler will require the strength of two men; knocking out a fireplace to take a room-heater and back boiler is quite heavy, dirty work; chopping through an exterior wall is not much fun and climbing on the roof to drop a flue liner down the chimney is not for the faint-hearted.

The electrical installation is a part that worries some people, but it is quite safe if carried out carefully and logically. However, if you are worried about this you can call in an electrician.

The part that causes most apprehension is the gas connection, and this *must* be left to an approved gas fitter – it is against the law for an unqualified person to interfere with the gas supply. This part of the installation is covered under 'Commissioning the boiler'. Some large suppliers of heating materials offer a package deal which includes checking the installation, making the gas connection and starting up the system. Area Gas Boards also offer this service for a set fee, or can put you in touch with an approved fitter.

When to do it

The ideal situation for installing central heating is undoubtedly an empty house where floorboards can be left up and no heavy furniture has to be moved; but this is unlikely unless one is just moving house. If possible, try to organize it so that you can do it during the summer months so that you will be under no pressure to get it working to keep warm, and any outside work such as putting in flue liners or cutting flue holes in walls will be more pleasant.

How long will it take?

If you reckon that half the cost of a professional installation would be for labour, it becomes clear that many hours of work are required to do it yourself. Reckon on between four and eight weeks working in your spare time on a three-bedroom house. Always reckon on it taking longer than you think.

Choice of fuel

The kind of fuel used to fire your boiler is in some ways a personal choice, but it does need very careful consideration. The wrong choice at this stage may mean changing your boiler in the future, and this would be very costly. The needs of your home and your lifestyle will dictate to a certain extent what type of boiler and fuel is chosen. All boilers come with very precise and detailed instructions, so you will not lack the information to install them correctly. It is not easy to make direct comparisons of fuel costs as different fuels are rated in different ways. To find the current costs of the different fuels, look for articles on home heating in newspapers and magazines.

Gas

The most popular choice of fuel for central-heating systems today, gas also has the largest and most varied range of boilers. There are no storage problems, it's clean and easy to control, but is not available everywhere. Gas boilers need servicing about once a year. A standing charge is made every quarter whether you use any gas or not. The advent of balanced-flue boilers has increased the versatility of this fuel because they can be sited on any external wall without requiring long sections of flue.

Solid fuel

Not as clean and controllable as gas, solid fuel requires storing in a shed or coal bunker since it must be kept dry for maximum efficiency – heat is lost drying off wet fuel and it can corrode the boiler.

Solid-fuel boilers generally operate continuously throughout the winter (although at a very low rate during the night), which has the advantage that the house fabric always has some residual heat. In cold weather, room-heaters require refuelling up to three times a day, hopper-fed boilers once a day.

Fuel can be bought more cheaply in the summer when special offers are available, and with a large stock of fuel you can be sure of your heating throughout the winter.

Oil

Like gas, oil is very controllable but, unlike gas, is available everywhere. A storage tank is necessary and so is access for tanker deliveries – tankers can weigh up to 20 tonnes and some roadways and drives may not be suitable. Installation costs will be more than for other systems and more frequent maintenance is required to keep the boiler operating efficiently. The storage tank should not be sited in an exposed position as oil thickens in very cold conditions and will not run easily through the feed pipe. The tank will require regular painting to prevent rust.

Some oil boilers are noisy and require siting in a boilerhouse. The storage of oil may affect your house insurance and some local authorities have very strict safeguards against the possible spillage of large quantities of oil. Oil boilers can produce very large outputs and are therefore a good choice for large houses.

Wood
Recent years have seen an increase in the availability of wood-burning stoves, many of which are capable of running a central-heating system.

Although burning wood looks a very attractive alternative to other fuels, certain factors have to be taken into consideration. If you live in a smokeless zone, it rules out wood burning on a regular basis. For such boilers to reach their quoted outputs, good-quality seasoned logs are required; you need space to store logs for up to a year under cover. Burning green and damp wood produces a tar-like substance that coats the chimney, can cause condensation and lack of draught in the flue, and can cause chimney fires.

Very little maintenance is required for these boilers apart from regular chimney sweeping. In a rural property with the prospect of a regular supply of logs and the space to store them, this type of boiler could be a very good buy.

Propane gas
A relative newcomer to the central-heating market, propane gas means that gas central heating can be considered even where mains gas is not available. A storage tank is required (usually leased from the supply company) and so is access for tanker deliveries. Since propane gas is an oil-based product its price is related to that of oil.

CHOICE AND SITING OF BOILER

Having looked at the choice of fuels, next consider the boilers themselves and the restrictions that your own home may place upon that choice.

Gas
If you are considering a free-standing or wall-hung gas boiler, a balanced-flue type offers the greatest choice of site and is easiest to install.

Providing you have the exterior wall space, and you observe the recommendations regarding windows and doors, a balanced-flue boiler can be sited almost anywhere in the house. Some balanced-flue types can even be sited on an interior wall adjacent to an exterior wall and the flue taken out on either side.

Remember to allow access for servicing according to the manufacturer's instructions. If you live in a terrace house the area of exterior wall may be limited, but you may have a chimney in the living room, in which case a room-heater with back boiler would be a good choice. The chimney should have a flue liner inserted to prevent combustion gases attacking the brickwork.

If you install a conventional-flued boiler, arrangements must be made for a permanent air supply, not just an occasionally opened window. A conventional-flued boiler must not be sited in a garage or workroom where inflammable materials or vapours may be present. A balanced-flue boiler is advisable in a small kitchen with an extractor fan fitted, because otherwise fumes could be drawn into the room.

With conventional-flued appliances, in order to get a good draught, chimneys should preferably rise straight up for at least two metres before any bends are made to take it outside. No bend should be more than 135 degrees.

Solid fuel
One of the factors affecting the layout of solid-fuel systems is that the hot-water cylinder (or at least one radiator if the hot water is not heated by the boiler) must be on a gravity circuit. This is because solid fuel will continue burning and supplying heat even if the water thermostat is turned right down and the pump turned off. Hot water in a solid-fuel system can be pumped, but special precautions are necessary. Since the pipework between the boiler and the hot-water cylinder will be large-bore, the boiler needs to be sited close to the hot-water cylinder or vice versa. Ideally the boiler should be below the hot-water cylinder. The horizontal distance between the two should be less than the vertical distance, and pipe runs should be as short as possible. All this may limit your choice of position for the boiler.

Free-standing boilers require a flue pipe and a good supply of fresh air to burn well. Solid-fuel boilers must not stand on a timber floor but on a non-combustible, solid base. Open fires with back boilers, and room-heaters, will be situated in a fireplace opening and the normal chimney will be satisfactory, providing it is in good condition.

Flexible stainless-steel liners are not suitable for solid-fuel appliances. The building regulations concerning hearths and fire surrounds must be observed.

Siting an AGA type cooker and boiler in a convenient place for cooking may mean finding a new site for the hot-water cylinder to keep pipe runs short.

Oil
Both balanced-flue and conventional-flue boilers are available for oil, and both are suitable for kitchen installation although a boilerhouse would be better. You should take into consideration the proximity of the oil tank to keep the feed pipe to a reasonable length. Oil storage tanks can be made to order.

HOW CENTRAL HEATING WORKS

Fully automatic gas central heating

Simplified sequence of operations

The heating system

The temperature falls in the house and an electrical contact is made in a temperature-sensitive **room thermostat** sending a signal to the controller.

The **controller** sends signals to the boiler, pump and heating zone valve.

The gas ignites in the boiler and heats up the water.

The **pump** pushes heated water around the system through the **zone valve** (now open) to the radiators. Heated water circulates through the radiators, pushing cold water in front of it.

The water, now cooling, returns to the boiler and is reheated and recirculated until the required room temperature is achieved.

The **room thermostat** senses the new, higher air temperature and an electrical contact is broken, sending a signal to the controller.

The **controller** sends signals to the boiler, pump and heating zone valve; the gas is shut off in the boiler, the pump stops and the zone valve closes.

The hot-water system

The **cylinder thermostat** fixed to the hot-water cylinder senses a fall in the temperature of the water as hot water is drawn off for a bath and cold water replaces it. An electrical contact is made and a signal is sent to the controller. The **controller** sends signals to the boiler, the pump and (this time) to the hot-water **zone valve**. The sequence follows as before except that heated water is circulated around the cylinder circuit. The system shuts down when the water in the cylinder reaches a pre-set limit.

Note When the heating and the hot water are both in demand, the controller opens both zone valves and will shut each valve independently as the required temperatures are achieved.

Feed and expansion tank

Fitted in the loft. Cold water is fed through a ball valve from main supply into tank. When the level falls, the ball valve opens and supplies fresh water. When the level rises, the ball valve shuts off supply. The tank makes up the losses in the system and also takes up the expansion of water due to heating.

Open vent

Provides an additional safety outlet should the boiler overheat because of a fault.

Safety valve

Fitted in the flow pipe, close to the boiler. Will open automatically when the pressure reaches a pre-set level.

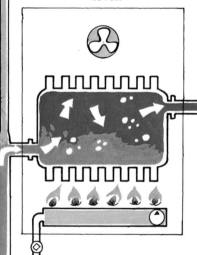

Balanced-flue boiler

Gas ignites automatically on a signal from the controller and heats up the water in the jacket. A fan extracts fumes through the wall duct to outside, and draws in fresh air for combustion.

Controller
The 'brain' of the system.
Incorporates a clock which can be set to switch the heating and hot water on and off automatically, according to the needs of the household. Some types incorporate a rechargeable battery which keeps the clock running during power-cuts, avoiding the need to reset.

Zone valve 2
The hot-water zone valve opens and closes by electric motor, according to the temperature of the water in the hot-water cylinder sensed by the cylinder thermostat, or when the controller clock comes on.

Pump
Pushes heated water through the pipework to radiators and the hot-water cylinder.

Room thermostat
Usually sited in the hall. Can be set to any temperature from 10° to 30°C, but most often set at 18°C. This control senses the temperature of the air around it and will switch on the heating when the air temperature falls below the setting on the dial during a 'heating on' period set on the controller.

Air vent
Has a knurled cap which can be unscrewed to release trapped air from what is often the highest part of a pumped system.

Hot-water cylinder
An indirect system which has an internal coil of pipe circulating hot water which heats the main body of water stored in the cylinder; when heated, this can be drawn off.

Cylinder thermostat
This is fixed by a strap to the outside of the hot-water cylinder (with no insulation between it and the cylinder) about one third up from the bottom. It senses the temperature of the water in the cylinder and will switch on the hot-water section of the system during a 'hot water on' period set on the controller.

Zone valve 1
The heating zone valve opens and closes by electric motor, according to the temperature of the air sensed by the room thermostat, or when the controller clock comes on.

Radiator
Each radiator has a wheel valve, which can be turned on or off by the householder, and a lockshield valve which is adjusted when the system is first put in use. This restricts the flow of water through the radiators nearest the boiler and allows those furthest away their share of the heat, thus balancing the system. An air-bleed valve is fitted into one top corner of the radiator and is opened with a radiator key to allow trapped air to escape and water to fill the radiator.

Gas boilers

Gas is the most popular choice of fuel for firing a central-heating system and this is reflected in the number and variety of boilers available. They fall mainly into three groups:

1 Gas room-heaters, sometimes called fireside boilers.

2 Floor-standing boilers.

3 Wall-hung boilers.

Flue types

Both floor-standing and wall-hung boilers are available in conventional-flue and balanced-flue models. Conventional-flue or open-flue boilers are the ones with the familiar large-bore pipe rising from the top to be vented high up, outside the house.

Balanced-flue boilers have grown in popularity over recent years because they have certain advantages over conventional types. Since they draw their air for combustion from outside the house and also exhaust the waste fumes outside, no special ventilation is necessary for the room in which they are sited. The balanced-flue outlets are

Above: The floor-standing, balanced-flue boiler fitted in the corner under the worktop in this kitchen shows how well a central-heating boiler blends into modern kitchen design.

Right: This slim floor-standing model with a conventional flue would look equally at home in a modern kitchen or with the more traditional units shown here.

Far right: Even the smallest kitchen can accommodate a boiler: this range-rated, wall-hung boiler with a balanced flue fits neatly alongside the top cupboards without spoiling the look of the kitchen, and is easily accessible.

small, they are supplied with the boiler and there is less work entailed than there would be in fitting an external flue pipe. There are regulations concerning their distance from openable windows and doors.

All boilers should be sited so that the servicing requirements stated in the manufacturer's instructions can be observed, and a maintenance contract either with the manufacturer or the local Gas Board arranged.

This balanced-flue, wall-hung boiler is range-rated and incorporates a cast-iron heat-exchanger.

A compact wall-hung boiler with a decorative finish fitted between dark oak top cupboards. Available in either a conventional-flue or balanced-flue version.

Wall-hung boilers

An increasing range of wall-hung boilers is available, mostly with balanced flues. New developments have reduced their weight and size and some are very small and light indeed. They can be fitted almost anywhere on an outside wall, providing the recommendations concerning the proximity of flues to windows and doors are observed. Some models can be fitted on an internal wall adjacent to an external wall with the flue coming out sideways; there are very few kitchens in which a gas boiler cannot be accommodated, and they can in certain circumstances be fitted in other rooms of the house.

Some models, usually low-water-content boilers, have copper heat-exchangers to reduce weight. Other models have robust and well-proven cast-iron heat-exchangers.

Selecting a boiler

The big central-heating suppliers will send lists (on request) of all boilers with full specifications.

Modern boilers are usually range rated, which means that the boiler output can be adjusted at the commissioning stage to match your system almost exactly, so long as it falls within the minimum and maximum output quoted for that particular model.

Solid-fuel boilers

Modern solid-fuel boilers are as efficient as the best gas or oil boilers. Most are now gravity fed, that is, the fuel — anthracite in a hopper above the fire-bed — is fed automatically by gravity to the fire, according to the burning rate.

A fan revives the fire when an increase in output is demanded by the boiler thermostat; when the desired water temperature is achieved, the fan switches off and allows the fuel to burn under natural draught conditions. Ash is formed into clinker, allowing dust-free removal and disposal.

A flue pipe will need to be fitted to carry the fumes clear of the house; alternatively, the boiler can be connected to a chimney.

The room in which the boiler is sited will require permanent ventilation to allow the fuel to burn properly. This will be specified in the manufacturer's instructions.

Right: A small, economical boiler which features a high fuel-holding capacity and external riddling for dust-free operation.

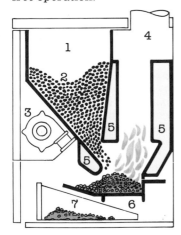

Hopper-fed solid-fuel boiler
1   Hopper
2   Fuel
3   Fan
4   Flue
5   Waterways
6   Grate
7   Ashpan

Oil boilers

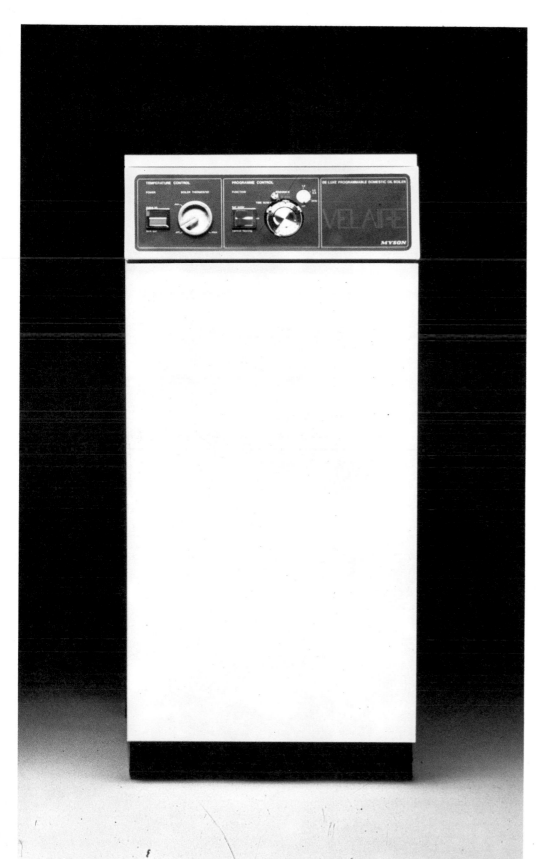

If you are choosing an oil-fired boiler, first consider where you will be siting the oil storage tank (see page 77).

There are two main types of oil boiler suitable for domestic installation — the pressure-jet and the wallflame.

Pressure-jet
Pressure-jet boilers work by atomizing the fuel for combustion by injecting oil and air under pressure, rather like a spray gun. Down-firing pressure-jet boilers are preferable to the standard pressure-jet and most of the newer domestic models are now of this type.

The larger pressure-jet boiler for bigger houses should be installed in its own boiler-room where its operating noise will not be noticed.

Wallflame
The wallflame oil boiler is quieter than the pressure-jet and it is the most popular where a kitchen is the only possible site. The oil is initially vapourized by an electric heating element but, once ignited, the process is continued by the boiler itself.

Oil boilers are ideal for fully automatic control and will do anything that a gas boiler can do. They are more expensive to install, however, mainly because of the need for a storage tank and its ancillary equipment.

Flues
Oil boilers require a different type of flue from gas boilers because much higher flue temperatures are reached. Consult the boiler manufacturer's instructions.

Left: A modern, high-efficiency oil boiler suitable for kitchen installation. This one has a built-in programmer.

Solid-fuel room-heaters

Automatic central heating with a boiler in the kitchen is a neat concept, but a living room needs a focal point; for hundreds of years this has been supplied by an open fire in the hearth. Solid-fuel appliance manufacturers have managed to combine the function of a living fire with a boiler to supply hot water and central heating. And the development of the back boiler has led to the modern room-heater.

Solid-fuel room-heaters require probably the least maintenance of any boiler type and, being robustly constructed (mainly in cast iron), they last virtually forever.

Maintenance is restricted to occasionally brushing out the soot, and sweeping the chimney twice a year. The disadvantages are mainly the removal and disposal of ashes and the slower response of solid fuel compared with gas or oil.

Room-heaters have waterway thermostats, and by setting a knob the air supply to the firebox is restricted to control the rate of burning and thus the temperature of the water leaving the boiler. When the temperature of the water falls, the thermostat opens the air supply again to increase the burning rate.

The fire is usually banked up at night and the thermostat turned to its lowest setting, to be revived in the morning by turning up the thermostat and riddling the grate.

Modern manufactured fuels are available in different forms for different purposes and can be burned freely in smokeless zones. Good-quality types ignite easily, give off a lot of heat and burn slowly to a fine ash to make removal easy. Some room-heaters will burn housecoal smokelessly, which makes them even more economical.

This inset room-heater provides space heating and hot water as well as central heating. It is one of the newer types of heater which burn house coal smokelessly. As house coal is cheaper than manufactured smokeless fuels, it is economical to run.

The unusual design of this room-heater means it can be used either as an open fire or, by lowering the closure plate, as an enclosed room-heater. It burns low-cost house coal without making smoke.

Below: Cutaway of a typical inset room-heater.

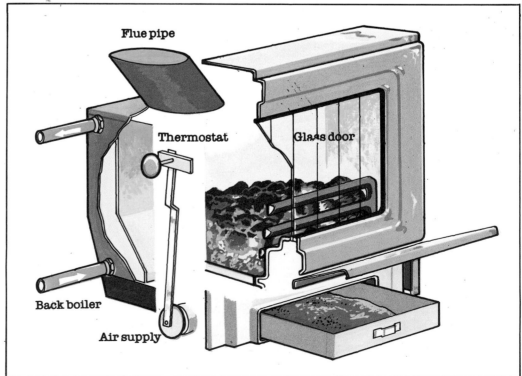

Flue pipe

Thermostat

Glass door

Back boiler

Air supply

This free-standing room-heater is available either as a space heater only, for a large room, or (with a high-output back boiler) for central heating and domestic hot water. This one has a traditional all cast-iron finish and many designs are available.

Free-standing room-heaters

The usual type of room-heater is designed to fit into the fire opening with only the front exposed to view. The actual back-boiler part of the fire is recessed into the chimney breast and the pipe connections made through the side. Another type is free-standing, and is usually fitted into a large recessed opening with a short length of flue pipe connected to the chimney. Some old houses have very attractive large open fireplaces (originally designed for burning logs) which, with a free-standing room-heater fitted in them, can be made useful while still retaining their old-fashioned character.

Open fires

Surprising as it may seem, some specially designed open fires are powerful enough to run central-heating systems; open to the atmosphere, they don't have quite the control that fully enclosed room-heaters have, but they look more attractive. Some open fires have underfloor ducting and an electric fan to control their burning rate. Some have very large ashpans and need emptying only twice a week.

These represent just a few of the huge range of solid-fuel appliances available today. There have been tremendous advances in the design of domestic heating boilers for coal in recent years: whereas the old-fashioned coal fire was extremely inefficient, most of the heat going up the chimney, modern solid-fuel heaters have efficiency figures of 60 or 70 percent. This country's huge supply of coal (estimated to last another 600 years at the present rate of use) means that it will play an important part in heating for many years.

A well-designed, modern free-standing room-heater with an attractive vitreous enamel finish.

This open fire with a specially designed high-output back boiler will supply up to six radiators and domestic hot water.

Gas room-heaters

Like solid-fuel room-heaters, gas-fired versions are a good way of combining the boiler with a radiant heat source in the living room. Unlike solid fuel, however, the back boiler is separate and can be controlled independently, so that the room can be warmed on cool summer evenings without switching on the whole system; the central heating can also be used without the fire.

It's a good idea to fit a radiator in the room in which the fire is situated to supply background warmth when the fire is not in use. Fitting a thermostatic radiator valve will automatically control the radiator and avoid overheating the room when the fire is used.

Room-heater installation is fairly straightforward and a flexible liner fitted in the chimney will provide the necessary flue. An approved terminal suitable for gas is required to finish off. The fireplace is a logical position for the boiler, as the pipes can be taken out through one side of the chimney breast, the pump can be easily concealed in a small cupboard, and where the pipes rise to the first floor a false wall can be constructed. Gas room-heaters are suitable for either pumped or gravity hot-water systems and any form of automatic control. The newer models have range-rated back boilers which means that they can be adjusted exactly to match your central-heating requirement at the commissioning stage.

There is a wide choice of decorative surrounds, including hardwood and antique finishes. The gas connection will be made by an approved fitter, and, providing there is a supply pipe close by, should present no problem. Extra cost will be incurred if a new supply pipe needs to be run.

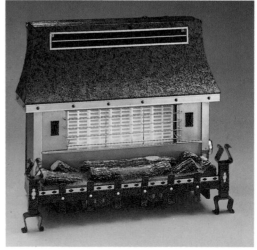

Above: Gas room-heaters are clean, neat and efficient. The boiler is hidden behind the fire and will supply all the heating requirements of an average house as well as domestic hot water.

Right: With a choice of two different boilers (both range-rated) and five different decorative surrounds, this gas room-heater would suit virtually any home.

Wood-fired boilers

There has been a proliferation of wood-burning stoves and boilers in recent years, as householders look for alternative fuel sources to cut their heating bills. Some of these are of British manufacture but a large percentage come from abroad, particularly Scandinavia where there is an unending supply of suitable wood.

The supply situation in your own area should be examined carefully, as logs need to be properly seasoned before they burn efficiently. You will require a covered area to store fuel, since burning wet logs is not good for boilers or flues. Wood cannot be burned in a smokeless zone.

Some wood-fired boilers can also burn other solid fuels, but unless your fire is designed for this it won't work properly because wood requires a solid bed to burn well, while coal needs a slotted grate.

Check the manufacturer's specification to see that the boiler will produce the sort of output you need from your central-heating system.

Flue requirements are similar to those for solid-fuel appliances – the manufacturer's installation instructions will give detailed recommendations.

Above: A wood-fired boiler which will also burn coal and smokeless fuels, giving it great versatility in areas where fuel supplies are uncertain. The glass door slides out of sight to give open-fire comfort.

Left: This fire comes from Scandinavia. The fire is made up from a kit of parts and not fitted into a fireplace like most boilers. It draws its air for combustion through an outside duct, thereby eliminating draughts.

Cooker/boilers

Evocative of vast farmhouse kitchens with a scrubbed pine table laden with pastries, the original AGA cooker was fired initially by wood, then by coal. Imitated by other manufacturers, the cooker/boiler is now available for all fuel types — gas, coal, oil, wood, even bottled gas and peat. The gas and oil types are also available in balanced-flue versions, making installation easier.

Ever since the AGA was invented in 1922, this type of cooker has been considered by culinary experts to produce the ideal conditions for cooking. Combining the cooker with the hot-water supply and then with central heating was a natural progression, and a large range of cooker/boilers is available from several manufacturers. Robustly constructed in cast iron, there is little to go wrong, and cooker/boilers have a long life expectancy.

Although wider than the average cooker, if this type of appliance includes a boiler it does not take up a great deal of space. Some models combine only domestic hot water with the cooking facility, so a separate boiler is required to provide central heating. This could be supplied by a small boiler which would shut down completely during the summer. By using a gravity primary pipe layout between the cooker/ boiler and the hot-water cylinder, year-round hot water as well as cooking would be provided economically. Water temperature is controlled by a thermostat on the cooker/ boiler.

Since it is not permitted to burn wood in a smokeless zone, if your area is or is likely to be designated one, choose your fuel carefully. Unless your cooker/boiler is a balanced-flue model it will also require permanent ventilation for combustion.

Top left: A modern version of the original AGA. This one is oil-fired and will provide domestic hot water as well as cooking facilities, but a separate boiler is needed for central heating.

Below left: This cooker/boiler will supply cooking, hot water and central heating for up to nine radiators. It will burn smokeless fuels, house coal, wood or peat.

Above: Primarily designed as a wood-burning appliance, this model will also burn coal or smokeless fuels as required. Newly developed, it features an electric element in the main oven for rapid response.

Left: A more compact solid-fuel cooker/boiler which provides hot water but not heating. It could be combined with a wall-hung gas boiler for heating only.

Panel radiators

Radiators are available in a variety of types and designs. Panel radiators are the most common and usually have fluted surfaces to give a larger heating area. Modern panel radiators are slim, efficient, elegant and easy to install. The air vent is usually already fitted and tappings at either end of the bottom are the only connections required. Where a higher heat output is required and space is limited, alternatives such as double, convector and fan convector radiators can be used. These will produce a higher output from a similar area.

Double radiators

These are simply two single panels, one in front of the other, with just two tappings for connection at the base. They do not actually give double the output, but heat emission is about 70 per cent higher than for a single-panel radiator of similar size.

Panel convectors

These are available either as single or double radiators. They have fins welded to the back, or in the case of a double convector, in between. This greatly increases the radiating surface and gives up to 40 per cent more heat than ordinary panel radiators.

Corrosion

It is most important to add a rust inhibitor to the central-heating system as soon as it is completed, because without it corrosion soon takes place and steel radiators are the first to go. If you own an older house with old-fashioned cast-iron radiators already installed, and you wish to retain these but update the rest of the system, special connectors will be required. (If you are going to dispose of them, remember that cast iron is valuable as scrap.)

A single-panel radiator looks elegant and unobtrusive when painted to match the decor of a room. The fluted surface gives a larger heating area from the same width.

Some types of double-panel radiators can be fitted with end and top grilles for neatness.

A towel radiator is the answer in the bathroom.

Types of panel radiator

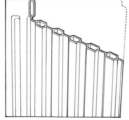
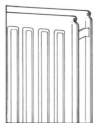

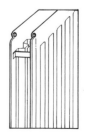

Single panel
Cutaway to show waterways

Double panel

Single convector
Viewed from the back

Double convector

Convector radiators

If lack of space is a problem, fan convector radiators are the answer, as their diverse shapes and sizes make them particularly useful in difficult situations. They can be installed instead of panel radiators throughout the house but, being much more sophisticated, are more expensive. They work by having air blown by a thermostatically controlled fan over finned heating pipes inside the casing. Convector radiators come with full fitting instructions, and are no more difficult to install than ordinary radiators, although each convector requires an electrical connection.

Fan convector radiators: warm air is blown through the grilles by a thermostatically controlled fan. The small convector in the foreground is for high-level installation (above a door perhaps) where it will blow down a curtain of warm air, and is ideal for kitchens.

A kickspace convector is a little more difficult to install but a good solution for a small kitchen. The thermostatic fan is controlled by a knob sited at a convenient point.

Skirting radiators

Skirting radiators blend into the decor of any room.

Choice of outlet grille position (top or front) gives this low convector versatility.

Skirting radiators can also be fitted throughout the house as an alternative to panel radiators. Again, they are more expensive, but are certainly less obtrusive than panel radiators and the heat is given off at floor level which is ideal. Relatively straightforward to fit, they work on the same principle as fan convector radiators and also come with full fitting instructions. If you object to large panel radiators in a room, skirting radiators are the answer.

23

Capillary fittings

Using capillary solder fittings is the most common way of connecting pipework in a central-heating system. These fittings are simple, effective, neat and cheap. Properly made, they become almost a part of the pipe itself and rarely give rise to problems. They are freely available in all the main small-bore central-heating sizes – 15mm, 22mm and 28mm – as well as a great number of other sizes. They cover almost every conceivable function in pipe connection, so you should have no difficulty finding fittings to make virtually every joint in your system, although compression fittings will need to be used where a joint has to be easily removable (such as on pumps and valves).

For extending older heating systems, adaptor couplings are available for joining metric pipe to the old imperial sizes.

To avoid confusion when buying a particular reducing tee, for example, make a rough drawing with the sizes of the outlets marked, and take it to your local heating supplier.

Integral solder-ring fittings are more convenient to work with than end-feed fittings and the price difference is not great.

Right: Capillary joints are simple to use and, properly made, are permanent and leakproof.

Far right: The slightly cheaper end-feed fittings require solder to be added after heating with the blowtorch.

Right: This type of capillary joint has an integral ring of solder set in a groove during manufacture. When the prepared and fluxed pipe ends and fittings are heated with a blowtorch the solder 'creeps' and seals the joint.

Solder-ring fittings

1 Straight coupling
2 Adapting coupling: imperial x metric
3 Reducing coupling
4 Female connector: metric x BSP
5 Male connector: metric x BSPT
6 Connector: copper x lead
7 Tank connector: metric x imperial
8 Tank connector: metric x BSP
9 Reducing piece: metric x metric
10 Adapting piece: imperial x metric
11 Female adaptor: metric x BSP
12 Male adaptor: metric x BSP
13 Adapting piece: metric x imperial
14 Elbow
15 Street elbow
16 Male elbow: metric x BSP
17 Female elbow: metric x BSPT
18 Back plate elbow: metric x BSP
19 Bath bend: metric x BSP
20 Overflow bend: metric x BSP
21 Equal tee
22 Reducing tee; branch reduced
23 Reducing tee; end reduced
24 Reducing tee; end and branch reduced
25 Reducing tee; both ends reduced
26 Equal tee: metric x BSP
27 Reducing tee: metric x BSP
28 Reducing tee: metric x BSP
29 Equal tee: metric x BSP
30 Air-vent bleed valve
31 Swept tee: metric
32 Reducing swept tee
33 End stop
34 Swivel tap connector: metric x BSP
35 Straight tap connector: metric x BSP
36 Straight meter union connector for gas
37 Bent tap connector: metric x BSP
38 Bent connector male: metric x BSP
39 Bent union connector female: metric x BSP
40 Bent union connector: male
41 Bent union connector: female for gas
42 Straight union adaptor: metric x BSP
43 Straight union connector: male x BSPT
44 Straight union connector: female x BSP
45 Male nipple (unequal): BSP x BSPT
46 Female nipple (unequal): BSP x BSP
47 Stopcock with compression ends

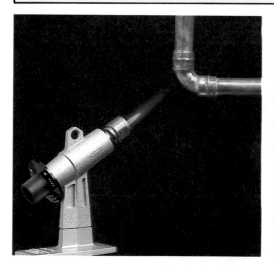

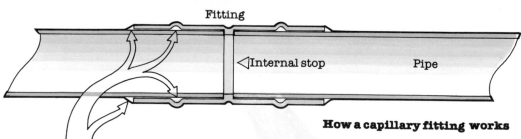

Fitting

Internal stop

Pipe

When heat is applied, solder runs from the integral ring and creeps to the end of the fitting to complete the seal

How a capillary fitting works

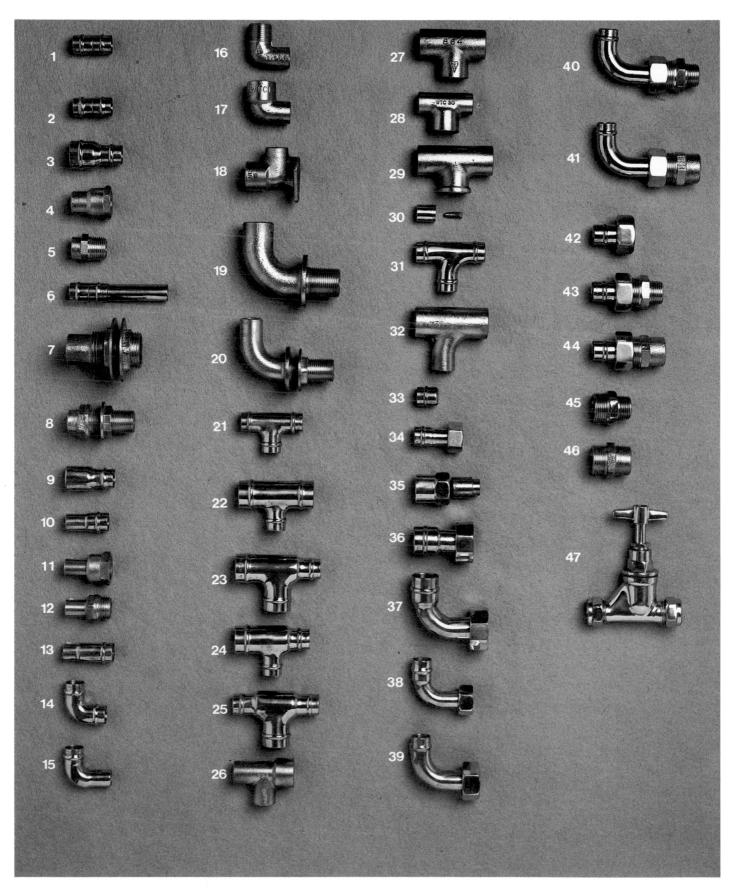

Compression fittings

Do-it-yourself plumbing consists basically of only two systems of pipe connection: capillary fittings (previously described) and compression fittings. Compression fittings are simple to use, and if care is taken in preparing the pipe ends excellent results can be expected. They are much bulkier than capillary fittings, however, and some people find them obtrusive where they are exposed to view.

Solidly constructed in brass, compression fittings are considerably more expensive than their capillary counterparts, and this will necessarily be a factor in deciding which type to use in your system. If, however, the prospect of using a blowlamp under the floorboards worries you, compression fittings are the answer. When planning

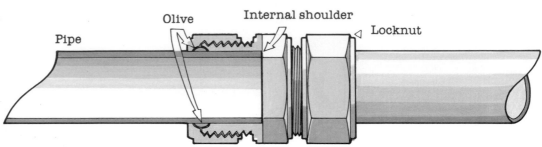

How a compression fitting works

When the locknut is tightened down, the chamfered internal edges of the fitting and the locknut compress the olive and, together with the jointing compound, form a watertight seal.

pipe runs, allow space for the manipulation of spanners to make the joint.

Another advantage is that where a joint is found to be leaking, after the installation is completed, it usually requires only tightening to

corrrect it, whereas on a capillary fitting the whole of the pipework must be drained before heat can be applied to effect a cure. Without a doubt, though, the biggest advantage of compression fittings over

Below: A selection of the more common types of compression fitting:
1 Gate valve
2 Radiator valve
3 Elbow
4 Air vent
5 Tee
6 Drain valve

capillary types is that the joint can be easily removed and remade, and it is for this reason that pumps and all types of valves have compression ends.

Note: If compression joints are remade, new olives (metal rings round the ends of the pipes) must be fitted – these are freely available in all sizes at plumber's merchants.

Compression fittings are available in the same variety of types and sizes as capillary fittings, so it should not be difficult to obtain all your requirements.

Below: A further selection of compression fittings:
1 Straight swivel coupling for connecting to a tank connector; a fibre washer is required at the point of connection
2 Elbow coupling
3 Equal tee
4 Bent swivel coupling for joining to a tank connector; a fibre washer is required at the point of connection
5 Wall plate elbow fitting
6 Straight coupling
7 Tank connector
8 Straight coupling for connecting a lead pipe to a copper pipe

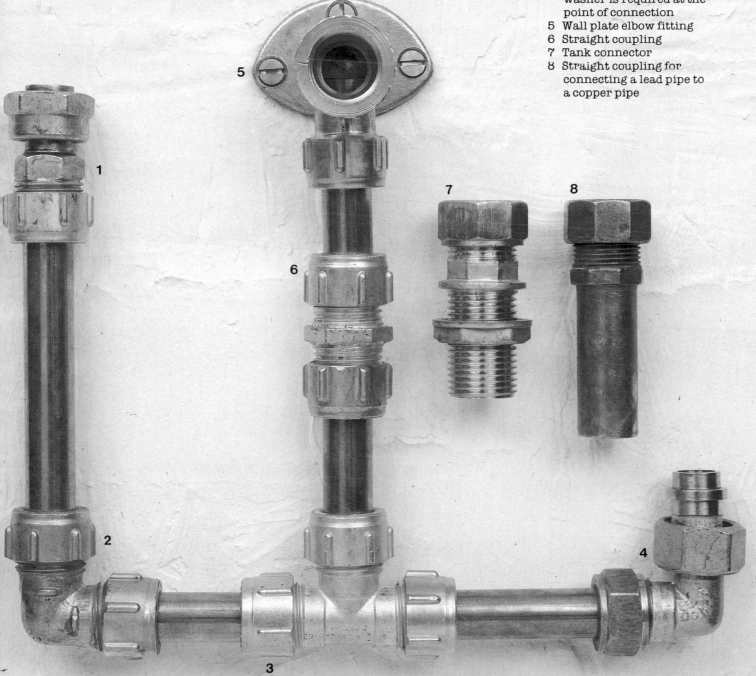

Pumps

Like other central-heating components, pumps have improved enormously in recent years, becoming smaller and more efficient as new materials are used for the moving parts. It was the invention of the small pump that revolutionized domestic central heating several decades ago. Before that time, wet heating systems depended on gravity for circulation, and ugly, large bore pipes were required to carry the flow of water through the system. Once pumps were incorporated, pipes as small as 12mm (½in.) diameter could be employed and the modern smallbore heating system came into common use.

The pump has to withstand being switched on and off many thousands of times a year, especially when incorporated into sophisticated control systems, but modern pumps are designed for this and even when used continuously their electrical consumption is modest.

Pumps need careful siting and fitting in the heating system and the instructions supplied with every pump should be followed. Some pumps have more than one speed as well as variable flow rates. This is to meet the demands of low-water-content boilers which require a minimum flow through the boiler at all times to avoid overheating. Some types can also have their moving parts replaced without removal from the pipework, so wherever you install your pump, make sure that there's sufficient room for maintenance.

It is important to ensure that the system is thoroughly flushed through before the pump is permanently fitted, and that an inhibitor is added to the feed and expansion tank when work is complete. Failure to observe one or both of these two simple rules is probably the main cause of premature pump failure. Apart from adding inhibitor annually, no maintenance is required for pumps and, if they are properly fitted, they will give many years of trouble-free service.

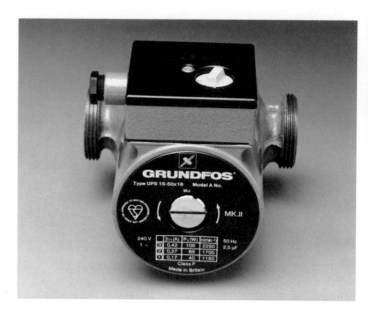

Above: A compact modern central-heating pump incorporating a facility to free the pump rotor without dismantling, should it stick. Pump performance is regulated by the white knob on top. The black plastic screw cap on the left is for the cable entry.

Below: A neat modern pump which would suit most installations. A three-speed regulator adjusts the performance over a wide range of applications.

A pump performance graph is supplied with every pump and will show at a glance the pump's performance at different regulator settings.

High head pumps

The increased use of lightweight, low-water-content boilers and microbore systems has created a demand for more powerful domestic central-heating pumps in order to overcome the higher frictional resistances imposed by these systems. If you are installing such a system you may not need a high head pump in a small compact layout, but check with your supplier which type you do need. Between them, the ordinary smallbore central-heating pump and the high head pump will cover just about any domestic installation.

A 6m head pump specifically designed for microbore installations. It incorporates a three-position regulator to cover a wide range of domestic installations.

Two-pump control system

Consisting of a special three-port controller pump, a conventional central-heating pump, a programmer, a room thermostat, a cylinder thermostat, check valve and isolating valves, this control system is unique in having two pumps as the basis for the separate control of heating and hot-water circuits.

In this system, the first pump (the controller) is fitted in the flow pipe from the boiler and supplies heat to the cylinder circuit via a branch pipe. It is controlled by a cylinder thermostat via the programmer and, when this circuit is satisfied, the controller pump switches itself off.

The second pump carries hot water to the heating circuit which is controlled by a room thermostat through the programmer. This pump circulates hot water throughout the heating circuit even when the controller pump is off.

Each pump is regulated to match the requirements of its particular circuit and the check valve stops unwanted gravity flow to the heating circuit when the domestic hot-water system alone is in operation.

Central-heating controls

Central-heating controls range from the boiler thermostat, which governs the temperature of the water leaving the boiler, to a system using zone valves to control heating and hot water independently through a central programmer. For an idea of how controls can be used, read the section on designing for control systems along with the practical installation instructions.

It is worth buying a complete control system from one manufacturer, as the wiring of the different components will then be much simpler. These are sold as 'plans' or 'systems', and detailed leaflets are readily available through heating suppliers.

The boiler thermostat
Used in conjunction with other controls, the boiler thermostat is usually turned to its maximum setting to act solely as an upper limit for water temperature.

The room thermostat
Room thermostats sense changes in the surrounding air temperature and will accordingly switch the pump either on or off to supply more or less heat to the whole house. For this reason they need to be mounted where they will not be influenced by factors other than the general house temperature. The hall is the usual place but it should not be fitted near the front door or over a radiator.

The cylinder thermostat
Sometimes called an aquastat, the cylinder thermostat is strapped to the hot-water cylinder and senses the temperature of the water inside it. It opens or closes a control valve to start or stop primary circulation through the cylinder as required.

A motorized zone valve will open or shut according to the demands from the thermostat controlling it.

A three-position motorized valve with a room thermostat and a cylinder thermostat will efficiently and economically control water temperature to the heating and hot-water circuits.

A modern controller featuring a digital read-out and two on-off periods every 24 hours for heating and hot water. An override button can change the mode of either.

A room thermostat senses the surrounding air temperature and reacts accordingly to control whole-house heating.

Simple to fit, various makes of thermostatic radiator valve are available. They give automatic local temperature control for rooms that become periodically overheated, such as kitchens and rooms receiving low winter sunshine. Under these circumstances they save heating costs by progressively shutting off the water flow to an individual radiator and returning to normal setting when the temperature falls.

A mechanical hot-water control valve reacts to the temperature of water in the cylinder.

A mini screw-adjusted valve is a good choice for balancing circuits or where a by-pass is fitted; once adjusted, it cannot be accidentally turned in the way that a conventional hand-operated valve can.

A check-valve will stop unwanted gravity circulation around a heating circuit when the pump is off and heat is only required at the cylinder via the gravity primary circuit. A simple weight in the valve is raised by the pressure created when the pump is activated and circulation restarted round the heating circuit.

The controller

The controller (or programmer) is basically a time-clock with the ability to switch on and off the control valves of the heating or hot water, the pump and the boiler, usually twice every 24 hours.

It can be set so that it switches the heating and/or the hot water on first thing in the morning, switches off when you leave for work, switches on again just before you get home in the evening, and finally switches off when you go to bed.

A simple override button can change the setting if required, for example at weekends if all-day heating is wanted.

Two- and three-way diverter valves

These valves, which are also known as priority or flowshare valves, will switch, or divert, the heating water from one circuit to another on receiving a signal from a thermostat (cylinder or room) via the controller.

Two-way valves can only divert either to the cylinder or heating circuit and priority is given to the cylinder circuit. This means that when hot water is required the heating circuit will cool. This is an economical way of controlling a system as the full boiler output is available for either the hot water or heating, and a smaller boiler can be specified.

Under some circumstances, however, the heating may not be all you desire. For example, if several people come to stay and the bathroom is busy, there may be plenty of hot water but inadequate heating.

The three-way diverter valve overcomes this deficiency since it can supply either the heating or the hot-water circuit, or both, as required.

Microbore

A microbore (or minibore) system is one using smaller pipes than the basic small bore sizes. Generally, a system using pipes down to a minimum of 15mm is considered small bore, whereas a microbore installation uses pipework of 10mm, 8mm and 6mm, which is something approaching the diameter of an electric flex.

Pipework

The tubing is soft copper which can easily be bent by hand, although bending springs (which fit over the pipe, not inside it) and special hand-benders are available. It is supplied by the coil in lengths of 10m, 20m, 25m, 30m, and 50m and not in straight lengths like small bore pipe.

The fact that it is so easy to manipulate and is available in long lengths means that it can be run easily through areas where small bore pipe would require a great deal of preparatory work. However, the very fact that it is so malleable means that it can easily be damaged or flattened by clumsy installation.

No tees or elbows are used in microbore pipe, but where a pipe needs to be joined couplings are available in compression, capillary solder-ring and end-feed types.

Manifolds

The main pipework is conventional 22mm or 28mm small bore pipe run to a central distribution point where the manifold is sited (usually one to each floor); from here the microbore tubing is connected and run to each radiator in turn.

The manifold is divided into two halves by a silver-soldered disc in the centre. One half carries all the flow connections and the other half carries all the return pipes.

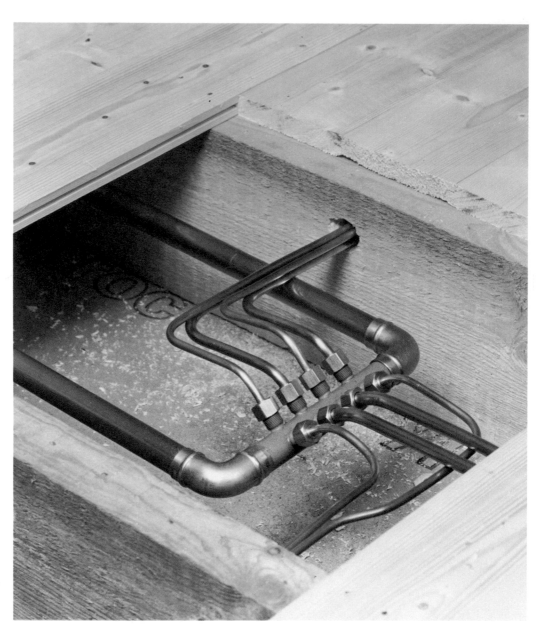

The heart of the microbore system is the manifold. Pipes of 22mm or 28mm carry the main flow and return to the central distribution point. From here pipes are run to each radiator using 10mm, 8mm or 6mm tubing, according to requirements.

A manifold connected to the main pipework, and ready to receive the microbore pipes.

The microbore pipes can be run through holes drilled in the joists or across small slots.

Microbore pipes are unobtrusive where radiators are fed from above in drop-feed systems.

Valves

Special radiator valves (double-entry valves) are used in a microbore system, although conventional ones with reducers can be used. Only one end of the radiator is used for both the flow and return, the other end being blanked off. Circulation through the radiator is achieved by fitting a length of 10mm pipe which passes in along the bottom.

Where a double radiator is fitted, a flexible insert is used instead to pass through the short entry and lie along the bottom of one panel. There are two types of double-entry valve, one with a hexagonal adjusting screw for individually balancing each radiator, the other a cheaper basic valve without this facility (balancing usually being done by the careful sizing of the microbore pipe to each radiator according to its output and the length of pipe required).

The microbore flow and return connections are standardized at 10mm in diameter. Where smaller tube sizes are required, reducing sets may be used to make all the connections.

Cutaway of a double-entry microbore valve incorporating a lockshield adjustment for individually balancing the radiator. The flow tapping is the one next to the radiator.

Right: A useful feature of these valves is that the handwheel section can be fitted in alternative positions, making connecting from above (for drop-feed systems) more direct; they can be angled towards a wall where pipes run along skirting boards

Cylinder conversion unit

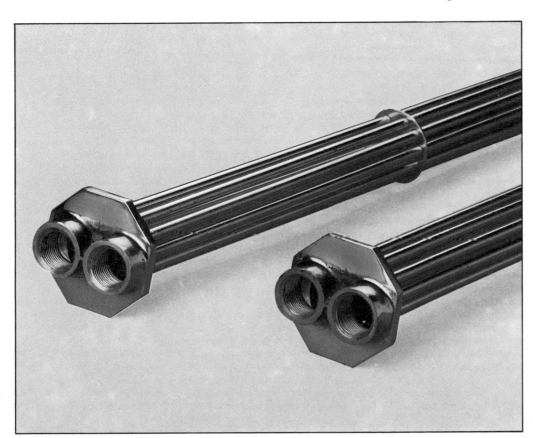

If a direct cylinder in good condition is available, a cylinder conversion unit can be used to convert it to an indirect cylinder at low cost. A conversion unit can be fitted in any system, not necessarily microbore. It is screwed into the immersion heater boss on the hot-water cylinder and connected via the primary pipework to the boiler. It can be used on both pumped and gravity systems, although a gravity system will take much longer to reheat the contents of the cylinder.

Two versions of the cylinder conversion unit. The longer 680mm type is screwed into the immersion heater boss, whilst the shorter 400mm unit is used for combination cylinders with a horizontal tapping. The primary pipework is connected via the two top tappings.

DESIGNING A SYSTEM

Central-heating design

Central-heating system design is not an exact science. There are too many uncontrollable factors involved which can upset the most careful calculating. For example, people stand and talk with the front door wide open; doors and windows can warp; kitchens can get unbearably hot during cooking; low winter sun can shine through windows, raising the temperature dramatically; gale-force winter winds can suck heat out of the house; and the number of clothes that can be hung on a radiator will insulate it more thoroughly than the jacket of a hot-water cylinder.

The advantage of designing your own central-heating system, providing it is done with care and is properly checked, is a system tailor-made for your home and your life-style. If you find the prospect of design daunting, most large DIY heating suppliers will design a system from details supplied by you. In either case you will need to know something about general house structure and to note down certain measurements.

General house structure

Walls: are they solid or cavity? If you don't know and the house exterior is brick, look at the way the bricks are laid. The diagram on the right shows a solid wall and a cavity wall.

If the outside walls are rendered then it is more difficult. Try to measure the thickness of the wall at a door or window. Deduct the thickness of the plaster and the rendering, and if it is 223mm (9in), which is one brick length, it is solid. If it is 280mm (11in), it is cavity. There are exceptions to this rule, and in some older properties thicker solid walls do exist. Houses built before the 1930s generally have solid walls. If you can't find out for certain, use the worst U value (the higher figure) in your calculations.

Roof: Note the type of roof structure and go up into the loft to see if there is felt under the tiles. You should also measure the thickness of the loft insulation. In passing, it's worth noting the change in U values (see page 38) when loft insulation is improved. Either now, or at a later stage, you could upgrade your insulation and base your calculations on the new improved figure.

Windows: Single or double-glazed? Metal or wood framed? When measuring windows, take the whole window area, not just the glass.

Ground floors: Solid or suspended? Some homes are a combination of the two – wood in the original structure and concrete for later additions.

You could also make a note of any conditions that particularly affect your house. Is it sheltered or exposed to strong winds? It is possible that one bedroom at the corner of the house takes the full force of the prevailing wind and gets no sun in winter, whereas another bedroom may be sited above a constantly warmed living room, thus gaining extra heat. Make an adjustment up or down of about 15 per cent on the total heat requirements for that room to compensate.

Another factor is the heat gained in some rooms from low winter sun shining through the windows. A neat solution to this problem is to fit thermostatic radiator valves to control the radiators individually.

Heat loss calculations

The purpose of making heat loss calculations is to define as accurately as possible the radiator sizes required to keep each room at its prescribed temperature when the ambient (outside) temperature is at -1°C. By adding up the output of all the radiators and allowing for domestic hot-water requirements the total output for the boiler can be calculated. This is simplified but indicates why these calculations are necessary.

First you will have to decide on the temperature required for each room in the house. There are standard recommendations for these and you should follow them as they have been calculated to be the best combination of comfort and economy.

To familiarize you with the procedure for working out the heat loss calculations for a room, we will take a hypothetical bedroom (shown opposite) and go through the calculations. For this you will require a pocket calculator.

The calculations are shown written into the heat-loss calculation chart (page 38) for simplicity. There are blank charts at the back of the book for working out your own heat losses.

First write the ambient design temperature (-1°C) into the box at the top of the table. Then decide on the temperature required for this room. By referring to the table of air changes and temperatures you can see that 16°C is the recommended temperature

Figures used in calculations

U Value	Tables of thermal transmittance or the rate at which heat passes through various building structures.
W/m2 °C	Watts per square metre by degrees centigrade.
7.44m²	This simply represents 7.44 square metres in area.
18.6m³	This figure represents 18.6 cubic metres in volume.
Air heat loss factor 0.33	A figure used in calculating the air changes in a particular room.
-1°C	This is the assumed outside air temperature used for all calculations. For most of the winter the ambient temperature will be well above this, but the system needs to be able to cope when the temperature does fall. If you live in a colder area you may take a figure of one or two degrees below this to give better design figures.
Temperature difference	This is the difference in degrees centigrade between the outside temperature (in most cases -1°C) and the required temperature of the room.

for a bedroom, so write this in the box marked 'Required room temperature'.

Exterior walls: This room has two exterior walls. Wall A is 2.50m x 2.40m = 6.00m².

Deduct the window area (1.25m²) from this, leaving an area of 4.75m². Referring to the table you can see that a U value of 2.1 is applicable to a 223mm (9in) solid wall. Write that in the table along with the temperature difference (the difference between the ambient and the required room temperature), and by working across the table you will arrive at a total heat loss figure of 169. This represents the heat loss

through this wall in watts.

Likewise write in the details for wall B but without deducting the window area.

Wall C is a partition wall to another bedroom, and so is not taken into our calculations. Wall D is a partition wall to the landing and so is not taken into our calculations either.

Ceiling: 3.10m x 2.50m gives a total area of 7.75m². The U value for a plaster ceiling with a pitched roof, tiles and felt and with 50mm of loft insulation is 0.6. The temperature difference is again 17°C, making a total heat loss through the ceiling of 79 watts.

The floor is not taken into our calculations because the room below is heated.

The window: The window is 1.20m x 1.04m giving an area of 1.25m². The U value for a single glazed wood-framed window is 4.3 and the temperature difference 17°C. The total heat lost through the window is therefore 91 watts.

These calculations represent the heat lost through the structure of the room. Some designers will go on to calculate the heat lost through party walls to the house next door, the heat lost or gained through interior walls, floors and ceilings in the same house heated to a slightly lower or higher degree. Such calculations are theoretical – interior doors in family houses are open much of

the time and there is a constant flow of air from one room to another.

There is one more calculation to complete the room losses – the air changes figure. This figure represents the heat lost from the room by draughts and opening doors. The table shows the air changes reckoned for each room type. A bedroom is reckoned to have one air change per hour.

The cubic capacity of the room is found by multiplying the wall height by the floor area. This is multiplied again by the air heat loss factor and again by the temperature difference.

$$7.75m^2 \times 2.40m^2 = 18.60m^3$$

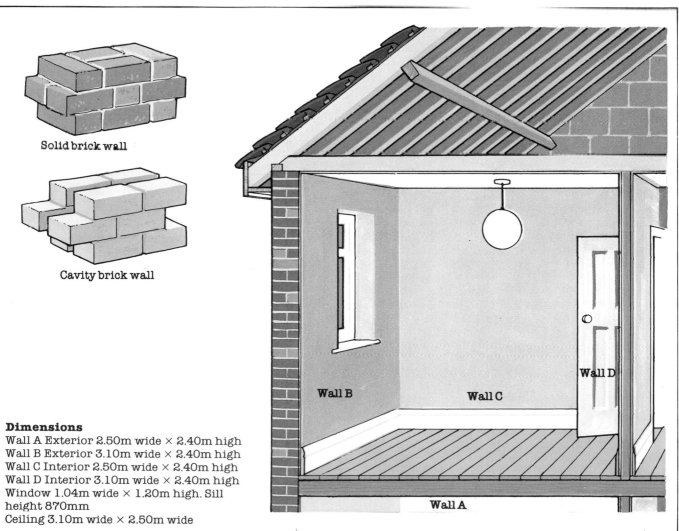

Solid brick wall

Cavity brick wall

Wall B

Wall C

Wall D

Wall A

Dimensions
Wall A Exterior 2.50m wide × 2.40m high
Wall B Exterior 3.10m wide × 2.40m high
Wall C Interior 2.50m wide × 2.40m high
Wall D Interior 3.10m wide × 2.40m high
Window 1.04m wide × 1.20m high. Sill height 870mm
Ceiling 3.10m wide × 2.50m wide

Air changes 1
×
Cubic capacity 18.60
×
Air heat loss factor 0.33
×
Temperature difference 17
= 104.34

The totals are now added together:
Exterior walls	435.54
Ceiling	79.05
Window	91.00
Air changes	104.34
	709.93

This figure is the watts per hour required to heat our hypothetical bedroom to the required temperature of 16°C, while the ambient temperature is at -1°C.

These calculations show how each room in the house should be treated.

You can now work through each room in your own home and note the details in the example room charts on page 104. When you have completed the calculations, write the totals into the house heat-loss calculator (in pencil in case of mistakes) on page 39.

Halls and landings with stairs can be arbitrarily separated into their upper and lower levels and the values added together.

In some rooms (single-storey extensions and all rooms in bungalows, for example), you will need to take both floor and ceiling areas into your calculations.

Table of U values

Structure	U value W/m² °C
Walls	
Cavity brick	1.5
Cavity brick with cavity insulation	0.5
Cavity brick/Lightweight block	1.2
Cavity brick/Lightweight block with cavity insulation	0.4
Solid brick 223mm (9")	2.1
Solid brick 105mm (4¼")	3.1
Doors (If glazed use U value for glass area)	
25mm (1") solid or cored (wood)	2.4
Roofs	
Pitched tile, loft space, plaster, uninsulated	2.2
Pitched tile, loft space, plaster, 25mm insulation	0.9
Pitched tile, loft space, plaster, 50mm insulation	0.6
Pitched tile, loft space, plaster, 100mm insulation	0.3
Flat asphalt, fibreboard, plasterboard, uninsulated	1.6
Flat asphalt, fibreboard, plasterboard, 50mm insulation	0.5
Flat asphalt, fibreboard, plasterboard, 75mm insulation	0.4
Windows	
Wood frame, single glazed	4.3
Wood frame, double glazed	2.5
Metal frame, single glazed	5.6
Metal frame, double glazed, no thermal break	3.2
Metal frame, double glazed with thermal break	3.7
Floors (reduce by 20% for thick carpet and underlay)	
Ventilated floorboards on joists	0.7
Solid concrete	0.8
Intermediate floors ; floorboards on joists	
Heat flow up	1.6
Heat flow down	1.4

If house is exposed to severe conditions add 10% to quoted U values

Add together the total for each room for the TOTAL HEAT REQUIREMENT for your house. There are two more calculations to be added to this: the domestic hot-water requirement and the boiler allowance.

Kilowatt/Btu conversion
To convert kilowatts to Btu's, multiply by 3412
To convert Btu's to kilowatts, divide by 3412
To convert kilowatts to watts, multiply by 1000

Heat loss calculation

	Walls	Length		Height (Width)		Area		Minus window area		Total		U value		Temp diff.		Heat loss	Total heat loss
Room Bedroom	A	2.50	×	2.40	=	6.0	−	1.25	=	4.75	×	2.1	×	17	=	169	Add all heat losses together
	B	3.10	×	2.40	=	7.44	−	−	=	7.44	×	2.1	×	17	=	266	
	C		×		=		−		=		×		×		=		
Required room temp. 16°C	Ceiling	3.10	×	2.50	=	7.75	−	−	=	7.75	×	0.6	×	17	=	79	
	Floor	−	×	−	=	−	−	−	=	−	×	−	×	−	=	−	
	Window	1.20	×	1.04	=	1.25	−	−	=	1.25	×	4.3	×	17	=	91	
			×		=		−		=		×		×		=		
Air changes	1	× Cubic capacity 18.60 × Air heat loss factor 0.33 × Temp. difference												17	=	104	709.93

Ambient design temperature − 1°C

Table of air changes and temperatures

Room	Air changes per hour	Temperature °C
Living room/ Bedsitting room	1	21
Dining room	2	21
Bedroom	1	16
Hall/Landing	2	16
Kitchen	2	16
Bathroom	2	19
Toilet/Cloakroom	1.5	16

1 The domestic hot-water requirement

Unless you already have efficient means for heating your domestic hot water (or are using a smaller type of room-heater with limited output) it makes sense to heat it from the same boiler that supplies the radiators. For most homes the boiler size does not need to be greatly increased as spare capacity is reckoned to be available because the outside design temperature of -1°C is not often reached. But if you have a large family and use a lot of hot water an allowance should be made.

The table above (right) shows common cylinder capacities and sizes and the heat allowances to be added to the central-heating totals if required.

2 The boiler allowance

There will be times when the outside temperature falls below the -1°C used in the calculations, so a percentage is added to the total so far calculated to cover this.
The usual figures are:
gas and oil boilers add 10-15 per cent;
solid-fuel boilers add 15- 25 per cent.
The higher figures for the solid-fuel boilers take account of the slower response of this fuel.

Gas and oil boilers run by programmers on intermittent switching should use the higher figure. The domestic hot-water allowance (if used) and the boiler allowance figures are now added to the total heat requirement and the final total is the heat

Hot water cylinders
Heat allowance 3 hour re-heat from cold

Capacity Litres (Galls)	Heat allowance kW	Size mm
117 (26)	2.5	840 × 460
		1070 × 380
140 (30)	3.0	910 × 460
		1520 × 300
162 (36)	3.5	1070 × 460

output (in watts) required for your house. Boiler outputs are rated in kilowatts, so divide your total wattage by 1,000 to give the total in kilowatts.

The choice of boiler
Having decided on the fuel type, you can now look through suppliers' lists to choose a suitable boiler. It is

unlikely that you will find one whose output exactly suits your figures, so choose the nearest available size. Using too large a boiler means it will be constantly switching itself on and off. Boilers work best at or near their maximum capacity. On the other hand, too small a boiler will not be able to cope in very cold weather.

Heat loss calculator

House structure (Refer to Table and write in type and relevant U value)		
	Type	U value
External walls		
Roof		
Floor		
Windows		

Heat loss requirements for each room	Watts	Room temp.	Sill height	Rad. size	Rad. output Watts
Living room 1					
Living room 2					
Dining room					
Kitchen					
Toilet/Cloakroom					
Utility room/Study					
Bathroom					
Bedroom 1					
Bedroom 2					
Bedroom 3					
Bedroom 4					
Hall/Landing					
Other					
Total heat requirements					
Domestic hot water. Cylinder capacity		litres			
Boiler allowance				%	
Total minimum boiler output				watts	
			Divide by 1,000 to give kilowatts		

Sizing radiators

In our hypothetical room, the height from the floor to the windowsill is 870mm. As this is where the radiator for the room is going to be sited, we need a radiator of a maximum height of 720mm, to allow 100mm clearance between the bottom of the radiator and the floor and 50mm between the top of the radiator and the windowsill.

Referring to the radiator sizing chart on this page, there is a range of radiators 685mm high. Look along the listed outputs in watts to find the figure closest to the calculated heat requirement for that room. The nearest figure in that size is 685 watts which will be a radiator 685mm high by 743mm wide. This is the way each room in the house is treated.

A very wide choice of radiators in different heights is available, as well as those shown here.

To make final calculations of your exact house heating requirements, use the amended radiator output figures and not the assessed room heat requirements, as these may well differ.

You should, of course, find out the make of radiator that your supplier offers, and obtain radiator emission charts for this particular type.

If you find any difficulty in obtaining the heat output from the radiator space available, consider using either a single convector, a double radiator or a double convector. If you still have problems, read the section *Radiators for difficult areas* on page 96. There will quite often be one room which causes problems, but with ingenuity and perhaps some compromise the solutions can usually be found.

Chart by
courtesy of Thorn/EMI

Height 380 mm (15″) — **Output in watts**

Length (mm)	Single	Double	Single convector	Double convector
743	407	689	522	934
946	515	873	663	1194
1149	623	1057	804	1455
1352	730	1240	944	1717
1556	837	1422	1085	1980
1759	943	1604	1225	2243
1962	1050	1785	1365	2506
2165	1154	1966	1505	2770
2368	1261	2147		
2572	1366	2327		
2775	1472	2507		
2978	1577	2687		

Height 530 mm (21″)

Length (mm)	Single	Double	Single convector	Double convector
540	393	657	502	889
743	536	899	689	1230
946	679	1139	875	1572
1149	820	1378	1061	1916
1352	961	1617	1247	2261
1556	1102	1855	1432	2606
1759	1242	2092	1617	2953
1962	1382	2328	1802	3300
2765	1521	2564	1986	3647
2368	1660	2800		
2572	1799	3035		
2775	1937	3270		
2978	2076	3505		

Height 685 mm (27″)

Length (mm)	Single	Double	Single convector	Double convector
540	501	832	641	1132
743	685	1138	879	1566
946	867	1443	1117	2001
1149	1048	1746	1334	2440
1352	1228	2048	1591	2879
1556	1407	2349	1827	3319
1759	1586	2649	2063	3760
1962	1765	2949	2299	4202
2165	1943	3248	2535	4644
2368	2120	3546		
2572	2297	3844		
2775	2474	4142		
2978	2651	4439		

Drawing plans

2 Plastic metric 500mm rule: Try to get a rule with one edge unbevelled for use with the set square.

3 Plastic set square: About 300mm (12in).

4 Protractor: A cheap protractor from any stationer's will do. Used in making isometric drawings.

5 Pocket calculator: Any cheap model with the basic functions will suffice.

6 Note pad: About 200mm x 125mm (8in x 5in).

7 Expanding rule: Metric and imperial.

8 Masking tape: For attaching paper to a drawing board or other flat surface.

9 Coloured pens: For marking pipe runs and other items on plans.

10 Fine-tip black marker or drawing pen.

11 Plastic rubber.

12 Pencil (HB or B): Keep it sharp.

1 Paper: The ideal choice is an A2 layout pad obtainable at art shops. Layout paper, while being white, has a good show-through which means that one drawing can be superimposed over another, making alterations and planning much easier.

Technique

1 Transfer all the measurements for one plane of the house to the paper, using the scale.

2 Line up the top edge of the set square with the top edge of the paper.

3 Place the unbevelled edge of the rule against the set square. Hold the rule firmly and slide the set square up and down the rule, drawing horizontal lines as you go.

4 Swing the rule at right angles and complete the ruling in the other plane.

Measuring the house

Starting with the ground floor and working from room to room, go round the house measuring all the internal wall lengths. The easiest way to do this is to make very rough freehand plans in the notebook and write in the measurements for each wall, or part of a wall, as you go.

Mark in doors and windows and also note and mark in any old hearths, since these can present an impenetrable area when making pipe runs. Also, mark in anything immovable that is likely to make installation difficult (such as baths, toilets and washbasins). Note which way the floorboards run.

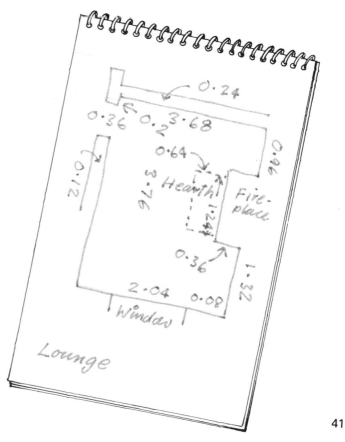

Finding a scale

Roughly add all your room lengths on one floor to find the largest dimension of your house and choose a scale that just fits comfortably on your paper. If, for example, the largest dimension of your house is 10.61 metres and you decide to make 50mm to 1 metre your scale, using the calculator, 50 x 10.61 = 530.5mm. A small dimension, say a wall at 250mm, will be 0.25 x 50 = 12.5mm. Work on as large a scale as you can − this will make details much easier to enter. When you have worked out a scale, write it in the top corner of your drawing for future reference.

Centring a drawing

Rule a line parallel to the edge of the paper and halfway across its width. Using the set square, rule another line halfway across its length at 90 degrees to the first line.

Halve the total dimension of the house and rule a vertical line parallel to the centre line at this measurement. This will be your starting line for all the other measurements.

Completing the plan

Work across the plan, transferring all your measurements as you go. Work lightly in pencil, using the rule and set square to keep all lines parallel. Include the interior wall thicknesses. When the dimensions one way are complete, work across the plan at right angles, keeping all the rooms square.

Once you are satisfied that your pencil drawing is correct, ink in all the wall outlines with a pen or a fine-point black marker. Rub out all the pencil lines. If you need to make corrections to any ink lines, use typewriter correcting fluid sold in small bottles with a brush in the top, (although coloured felt-

tips will sometimes bleed through it). You should now have a drawing looking something like the one below. You will need a drawing like this for each floor of your house.

You will find floor plans like this very useful for other DIY projects, such as carpet estimating or even planning extensions, so they are worth doing accurately and keeping for future reference.

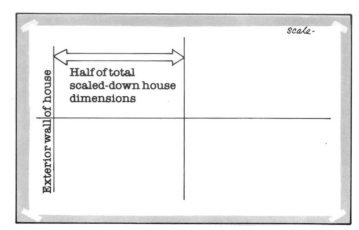

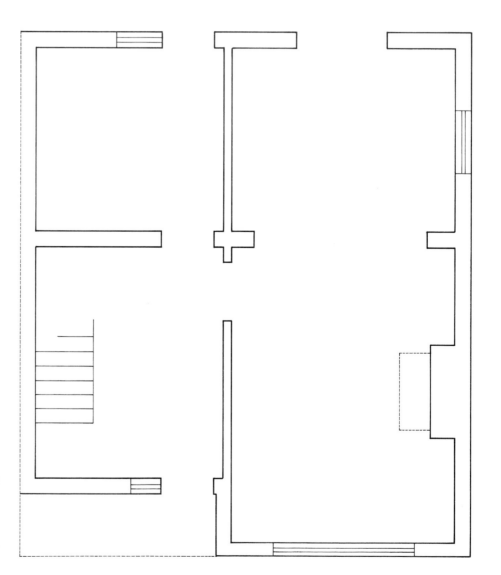

When planning pipe runs there are certain basic points that should be considered at the design stage. The first two listed below – the open vent and the feed and expansion pipe – must be incorporated into any design layout.

1 The open vent

All systems (except sealed systems which are not covered in this book) *must* have an open vent rising directly from the highest boiler flow tapping, and terminating over the feed and expansion tank. It should be made in a minimum pipe size of 22mm and should never be obstructed by any kind of control or valve. The safety valve, which should always be fitted, can be mounted on this pipe or directly on the boiler. In some circumstances, the heating flow pipe can be taken off the open vent, and in a gravity primary layout the flow to the hot-water cylinder is generally incorporated.

2 The feed and expansion pipe

Like the open vent, this should have an unobstructed flow to the boiler. Again, no valves or controls (except a drain valve) should be fitted between the feed and expansion tank and the boiler. This pipe supplies the boiler with water (if needed), and also takes up the expansion of the water by heating.

3 Gravity primaries

Gravity primary flow and return pipes should be kept as short as possible, and any horizontal runs of pipe should never fall from the boiler, but preferably show a rise. For every metre of horizontal pipe there should be at least half a metre of vertical pipe and, in any case, a minimum height between the boiler flow

tapping and the cylinder flow connection of half a metre. In most installations this is easily achieved.

4 Drain valves

A drain valve should be fitted in the lowest part of the pipework and wherever there are dropped loops of pipe. For example, one is needed at the boiler in the microbore layout on page 57 because the heating pipes rise to the first floor before dropping once again to the ground-floor heating circuit.

5 Air valves

An air valve should be fitted at each high point on a heating circuit (these are usually the radiators which incorporate air valves anyway) and the highest part of a pumped hot-water circuit. On a gravity circuit this function is supplied by the open vent.

You can now draw the radiators in position and roughly to scale. If their positions have been finalized you can draw in the boiler and the hot-water cylinder.

To demonstrate how to route pipes and position boilers and cylinders, we will work through a hypothetical semi-detached house and a small terrace house shown in the following pages. In our examples, both kitchens have a solid floor.

Pipe runs should be kept as short and direct as possible. For example, in upper-floor rooms it will be necessary to run pipes along joists or at right angles to them; but on ground floors you can run pipes under joists more directly at any angle. Draw pipes very lightly in pencil initially, in case you decide on a change of route. It is preferable, but not critical, to make the flow connection to each radiator at the coldest end. Thus, at radiator E the flow pipe (the

hottest) is towards the most exposed corner of the house, whereas the flow connection to radiator D is closest to the front door where there will be draughts.

Boiler position

We are using a floor-standing balanced-flue gas boiler (A) with pumped primaries, and the only real choice of position is along the back kitchen wall. The flue outlet will need to be kept clear of windows. It will also need to have a guard fitted because it is at a low level. A wall-hung boiler could of course be used here instead if floor space were limited.

The downstairs radiators

In general, fit radiators on exterior walls and under windows wherever this is possible.

A small radiator (B) is required in the kitchen. Fitting one by the door is logical as this space cannot be used for cupboards and it means that all the heating is on, or near, an exterior wall. (A certain amount of heat is of course given off by the boiler itself.)

The best position for the lounge radiator (E) is under the large window, and another radiator (C) is fitted in the dining area of the lounge under the small window. There will be a certain amount of cold air around the French windows, and an alternative arrangement would be to fit two small radiators on either side of these.

The hall radiator (D) is shown fitted under the window close to the front door where it will counter draughts.

The upstairs radiators

Radiator positions are fairly straightforward here. The only one that presents any problem is radiator F in the

bathroom – because of the WC and handbasin, the exterior wall is out of the question, and so siting it next to the bath is a good solution since it will provide heat where it is most appreciated. A towel rail can be clipped to the radiator.

The small landing gains heat from the hall radiator, but if another were necessary the logical position for an additional one would be at wall Y.

The pipe layout

We want to have this particular installation fully pumped, meaning a pumped cylinder circuit (instead of gravity) with two motorized valves to control heating and hot water. The hot-water cylinder, being where it is, dictates that you run the main flow and return pipes to the cylinder area, before you can take off the heating pipework. This is because the hot water must be circulated to the cylinder area (where the zone valves will be sited) from where either the central heating or the hot-water cylinder can be controlled independently without affecting the other circuit.

To find the best place to bring the flow and return boiler pipes up to the first floor, it is best to work out the vertical pipe layout first.

Using the same scale, draw the kitchen, bathroom and loft. Include the toilet and hand-basin because these will affect the run of the pipework; also show the hot-water cylinder and the wall of the airing cupboard.

Observing the major safety rule that an open vent must rise vertically from the boiler without obstruction, and that the cold feed must run directly to the boiler from the feed and expansion tank (also without obstruction), plan the vertical rise of these two pipes in conjunction with the flow and return to the control area as shown.

Right: When drawing plans of your proposed central-heating layout, as shown in these diagrams of a hypothetical semi-detached house, it is important to measure your plans very accurately. First make a floor plan of the ground floor, indicating the direction of floorboards and other relevant details, and marking radiator and boiler positions. Then draw the pipes, a different colour for the flow and the return.

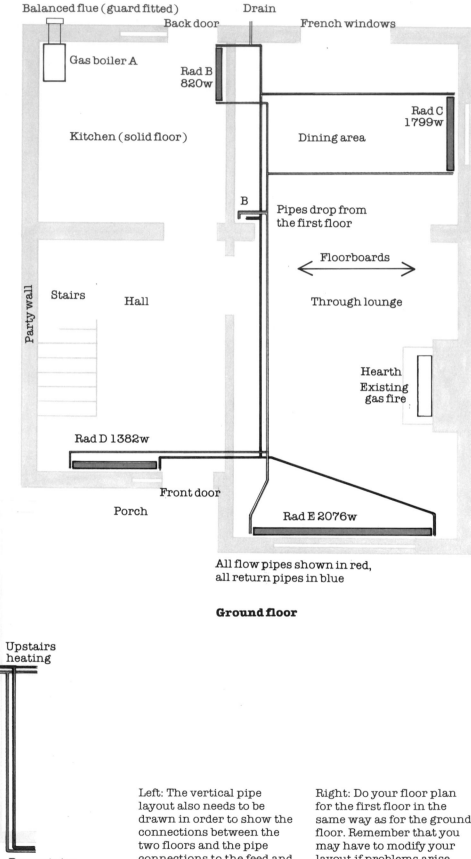

Balanced flue (guard fitted)

Drain

Back door

French windows

Gas boiler A

Rad B 820w

Rad C 1799w

Kitchen (solid floor)

Dining area

B

Pipes drop from the first floor

Floorboards

Through lounge

Party wall

Stairs

Hall

Hearth Existing gas fire

Rad D 1382w

Open vent

Loft

Feed and expansion tank

Front door

Porch

Rad E 2076w

Bathroom

Airing cupboard
Air vent

HWC

Upstairs heating

All flow pipes shown in red, all return pipes in blue

Ground floor

Pump

Kitchen

Boiler

Downstairs heating

Left: The vertical pipe layout also needs to be drawn in order to show the connections between the two floors and the pipe connections to the feed and expansion tank in the loft.

Right: Do your floor plan for the first floor in the same way as for the ground floor. Remember that you may have to modify your layout if problems arise when work is in progress.

The flow and return pipes, incorporating the open vent and the feed and expansion pipe, will rise from the boiler in the corner of the kitchen where they can be concealed behind kitchen units. Make a branch for the flow and return under the bathroom floorboards. By removing a section of floorboard in the airing cupboard you can site the pump and both zone valves conveniently where they will be accessible for servicing, and hidden from view. A false floor could be made to cover the pump and the heating zone valve to keep the whole installation neat. This will also simplify the wiring, as the pump, the zone valves and the cylinder thermostat will all be grouped together, and the boiler is not far away either.

Branch off conveniently to the hot-water cylinder and complete the circuit back to the return pipe. An air vent is incorporated at the top of the flow pipe to the cylinder because this is the highest part of this circuit. The hot-water zone valve can be fitted in either the flow or return pipe.

The heating zone valve can be fitted in the flow pipe just after the point where the branch is taken off for the cylinder circuit. This way, either circuit can be shut off without affecting the circulation of the other one.

The heating circuits
The flow and return now continue around the heating circuit but have to get down to the ground floor to complete it.

Looking at the house plan, you need to find a convenient and unobtrusive place to drop the pipes. There are various possibilities but probably the best place (to keep the pipe runs short) is just behind the stub wall where the through lounge was made (marked B). If necessary, the two pipes could be concealed behind a false panel.

The ground-floor pipe layout
Where the flow pipe is brought down, you can tee off under the floorboards — one branch going to the two front radiators, and the other to the two back radiators. We've managed to avoid running pipes through the concrete floor of the kitchen, only needing to chop out a short section at the flow and return connections to the kitchen radiator where the pipes come through from the living room.

The upstairs pipe layout
From the plan you see that pipe runs are relatively straightforward here, and keeping the flow and return close together will reduce the number of floorboards to be removed.

You can see from this that several attempts at the pipe layout may be necessary and you may even have to modify your layout at the installation stage when new obstructions are revealed by lifting floorboards. A good example of this is a sleeper wall under the floorboards at the junction of the two rooms where the lounge was made — you may need to locate a suitable gap in this.

Pipes rise from the boiler

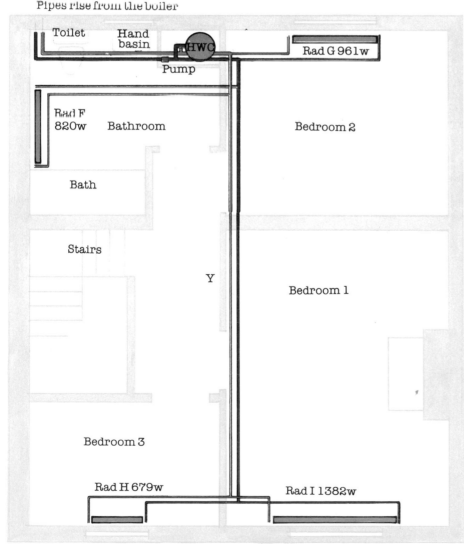

First floor

Making an isometric drawing

Although an isometric drawing looks quite complicated, it is not very difficult and one drawn to a large scale will show everything in your installation clearly.

To make an isometric drawing you will still need to draw floor plans as they make initial planning easier. Floor plans will also be useful in the future for estimating carpet requirements and other interior decorating projects.

Working in pencil, draw a long straight line close to, and parallel with, the bottom of the paper. Find the centre of this line and draw a vertical line up from it with a set square. All vertical lines will be made parallel to this. Place the centre of the protractor where these two lines bisect; mark off 20 degrees at the left hand side and draw a long straight line through the centre line and the 20 degree mark. This line represents the party wall at floor level. From now on, all lines in this plane are drawn parallel to this.

Do the same to the right hand side, marking off 20 degrees. All lines in the other plane are drawn parallel to this one. The line represents the front of the house at floor level.

Work across the drawing, transferring the measurements as you go. Start with the ground floor and work round until you have a complete outline of it, including door openings, chimney breasts, bay windows, and so on.

Work lightly in pencil, keeping the structural detail as simple as possible to avoid confusion. When the pencil drawing is complete for one floor, ink it in with a fine-point pen and rub out the pencil lines.

Allow sufficient room between the ground-floor and the first-floor plan, otherwise detail becomes confusing. Indicate the highest ground-floor radiator lightly before drawing the lowest part of the first floor to check that they do not coincide. This will give you the floor level for the next storey. Complete the inking of the second-floor area and only the part of the loft area that is relevant to your needs. Do not ink in any vertical structural lines. At this stage you should have a drawing looking something like the one shown here. (This shows only the ground-floor area.)

First draw the boiler

(roughly to scale) in position, then draw in all the radiators and the hot-water cylinder (draw the curves round a coin or other round object). Draw the feed and expansion tank in position and the part of the rising main approximately where you intend to break into it. Ink in these items and rub out the pencil marks.

Using your floor pipework plan as a guide, draw in the pipe runs and connect them to their various points. Keep

parallel runs of pipework a little away from each other for clarity. Mark all flow pipes in red and all return pipes in blue; this will show clearly which have to be connected to which. Mark all pipes with arrows showing the direction of flow.

At this stage, all your pipework, radiators, tank and boiler positions, and all connections should be complete, and your drawing will look like the one shown on page 48.

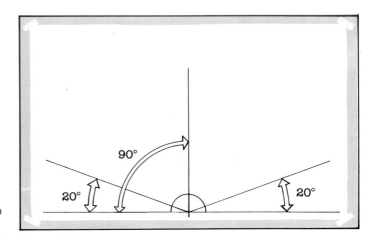

Above: The first stage in making an isometric drawing involves establishing the angles of walls in relation to the vertical.

Below: Complete the outline of a floor by drawing all walls parallel to one 20° angle or the other, noting your measurements.

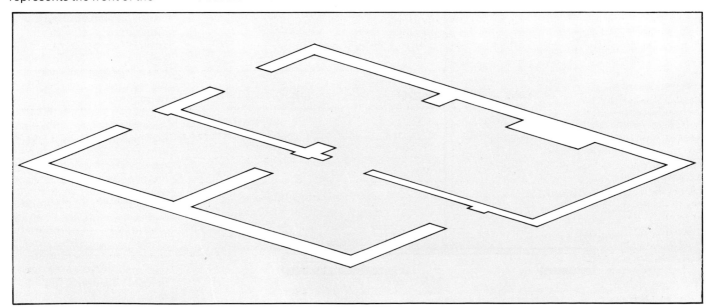

Sizing the pipes

The next step is to size the pipes. This is necessary because although the typical small bore system is based on 15mm pipes, there is a limit to how much heat a length of 15mm pipe can carry. For this reason pipe sizes must be increased above certain loadings.

There are standard recommendations for the performance of water through central-heating systems:

Water temperature The maximum temperature of water at the boiler is reckoned at 82°C; if it is any hotter the radiator surfaces become too hot to touch, if any cooler the system will be unable to cope with very cold conditions. This temperature is controlled by the boiler thermostat and 82°C is usually equivalent to the maximum setting.

Temperature drop The temperature drop (11°C) represents the difference in temperature between the flow from the boiler and the return through the circuit back to the boiler. This has been found to be the ideal for a fairly even emission of heat throughout the radiator circuit and a reasonably quick reheat at the boiler.

Maximum flow rate Excessive velocity of water through pipes can set up noises and vibration and high resistances due to friction. The flow through small bore pipework has been set at a maximum of 0.9m per second. Complicated calculations are used to work out the exact flow rate for a given section of pipework, but for domestic central heating, reference to the table (right) will supply all the information required. The length limits for small bore pipework are shown but it is unlikely in any domestic system that you will need such lengths.

To work out the pipe sizes, look again at the upstairs floor plan for our

hypothetical house overleaf, starting at the radiator furthest from the boiler – in this case, radiator H in the small bedroom.

The radiator output is 679 watts, well within the limits for 15mm pipe; so from radiator H to the tees that it shares with radiator I, mark 15mm on the drawing.

The connections from radiator I to the tees will also be in 15mm; this radiator is carrying 1382 watts.

Adding the two outputs together makes 2061 watts (which is still within the limit), so indicate 15mm pipe back to the downstairs connecting tees.

Radiator F at 820 watts

back to the tees on the main flow and return will be all right in 15mm pipe. Similarly, radiator G at 961 watts back to the tees near the hot-water cylinder is in 15mm pipe.

On the downstairs circuit, start with radiator D at 1382 watts. This is no problem, so 15mm pipe is marked to the shared tees with radiator E. Radiator E has 2076 watts so 15mm pipe is indicated. Adding the two outputs together makes 3458 watts, still within the limits, so mark 15mm pipe back to the tees at the dropping flow and return pipes. Radiators B and C at 820 and 1799 watts respectively are all

right in 15mm pipe even when added together (2619 watts). The whole of the downstairs circuit will be in 15mm pipe; but if you add all the downstairs radiator outputs together (1382 plus 2076, plus 820, plus 1799) it makes 6077 watts – above the limit for 15mm pipe. But since the downstairs circuit is in two loops, both within the limit, there is no need to increase to 22mm until the ground-floor tees of the dropping flow and return pipes.

Returning to the upstairs circuit, you have 22mm pipes at the tees for the downstairs and you can now work back towards the

Summary of pipe sizing   From radiator F back to boiler

Radiator/output (w)	Combined outputs (w)		Pipe size
F 2076	} 3458 (to rising tees)		15
D 1382			15
B 820	} 2619 (to rising tees)		15
C 1799			
Dropping flow and return pipes between floors plus rads. EDBC	6077		22
H 679	} 2061 (to rising tees)		15
I 1382			
F 820	to main flow and return		15
G 961	to main flow and return		15
Main flow and return pipes to HWC tees plus all radiators 9919			22
Flow and return pipes to boiler – all heating plus HWC at 3000 = 12919			22

Pipe sizing calculator

Boiler flow temperature			82°C
Temperature drop			11°C
Maximum flow rate			0.9m/s
Pipe size (mm)	Maximum heat load (watts)	Maximum length (m)	
15	6,000	24	
22	13,500	36	
28	23,800	175	

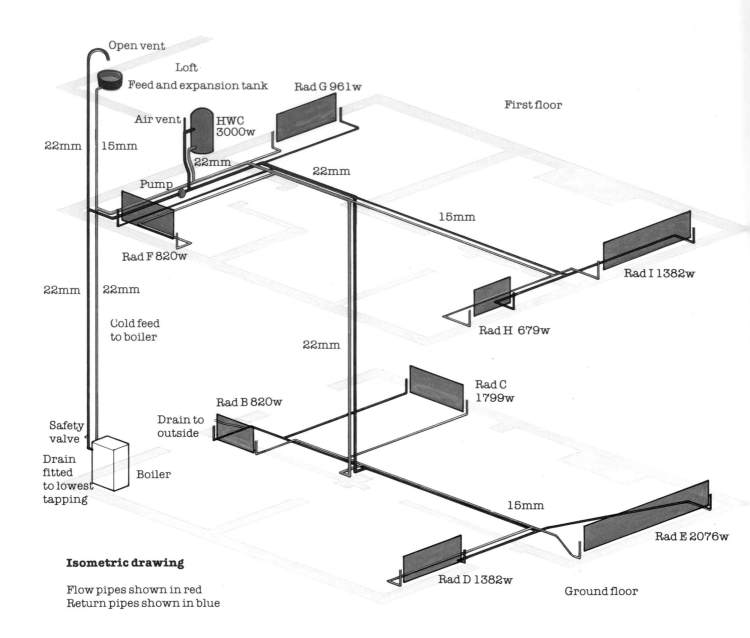

Open vent

Loft
Feed and expansion tank

Rad G 961w

First floor

Air vent

HWC 3000w

22mm   15mm

22mm

22mm

Pump

Rad F 820w

15mm

Rad I 1382w

Rad H 679w

22mm   22mm

Cold feed
to boiler

22mm

Rad C 1799w

Rad B 820w

Drain to
outside

Safety
valve

Drain
fitted
to lowest
tapping

Boiler

15mm

Rad E 2076w

Isometric drawing

Flow pipes shown in red
Return pipes shown in blue

Rad D 1382w

Ground floor

boiler. Add the heating load from the two radiators, H and I, to the total downstairs circuit – 6077 plus 679, plus 1382, making 8138 watts. Working back to where radiators F and G connect to the main pipework, add their heating loads to the total so far – 820 plus 961, plus the running total of 8138 watts makes 9919, still well below the maximum load for 22mm pipe. Indicate this for the pipework back to the hot-water cylinder tees.

Make an allowance of 3000 watts for the hot-water cylinder. This can be treated as another radiator as it is on a pumped circuit and could be connected in 15mm pipe; but because the pipe run is short (a long run would use a lot more of the costly 22mm pipe), and because the cylinder connections are large anyway and the zone valves have 22mm connections, it would be simpler to make the whole cylinder circuit in

22mm pipe.

All downstairs heating, all upstairs heating, and the hot-water cylinder load are now added together: 6077 plus 3842, plus 3000 make 12,919. This falls below the maximum heat load of 13,500 for 22mm pipe so it is safe to continue back to the boiler in this size.

Finally, the open vent must be in a minimum pipe size of 22mm, so continue in that size to the boiler flow tapping. The cold feed from

the feed and expansion tank to where it joins the main return pipe is made in 15mm pipe, the main return continuing in 22mm pipe to the lowest boiler tapping.

That completes the pipe sizing for our hypothetical layout and this is how you should treat your own design, working out heat loadings on the pipes and marking pipe sizes on your drawing, both for reference and for estimating materials.

The index circuit

Terms used in calculating pump performance

The index circuit: the particular section in the central-heating system that imposes the greatest resistance to the action of the pump.

Head, head loss: this refers to the pressure loss in the system that the pump has to overcome. Directly applicable to the index circuit.

Newtons per square metre (N/m²) a measurement of resistance. In imperial terms: inches or feet water gauge.

Pipe – actual length: the length of pipework measured directly from the plan.

Pipe – equivalent length: the actual length plus an allowance of 30 per cent to cover resistance through fittings.

Temperature drop: the difference in the flow and return temperatures throughout the system, here 11°C.

Equivalent heat flow: the sum of the heating load divided by the temperature drop. Rated in watts.

Flow rate: the rate at which water at different temperatures flows through pipes of different diameters.

Pressure loss: the drop in pressure caused by varying resistances throughout the circuit.

Mass flow rate: the total flow of water (throughout the system) that the pump

has to move. Rated in kilograms per second (kg/s). Imperial term: gallons per minute.

Specific heat: in the case of water this is the amount of heat required to raise the temperature of 1 kg through 1°C, and amounts to 4180 J/kg C (usually rounded up to 4200).

To work out the head loss of a particular system to apply it to a particular pump requires many tedious and complicated calculations and is not absolutely necessary. In addition, you are working with theory when looking at your plan but the index circuit may need modifying because of unforeseen obstructions when it comes to installation, so your calculations could be thrown off course anyway. If you wish to persevere the procedure is as follows.

To find the theoretical pump setting it is necessary to find and measure the circuit which presents the greatest resistance to the pump. This resistance is caused by long pipe runs or large heating loads (usually both). If the pump is set to overcome this resistance, lesser resistances in other parts of the system will be easily dealt with. The practical way of setting the pump regulator is fully explained

on page 95. Pressure is measured in Newtons per square metre (N/m²).

Most heating systems in two-storey houses are divided into two distinct circuits. The index circuit can often be seen simply by looking at the isometric drawing. If your design doesn't fall into this category then it may be necessary to work through more than one likely circuit and compare them to find the index circuit.

As an exercise in calculating the index circuit we will return to our hypothetical house and the system we designed for it. The drawing below shows the index circuit extracted for clarity from the complete isometric drawing. The way to calculate the resistances on this circuit is summarized in the table shown on page 51 and demonstrated in detail here.

1 Only the pipework directly in the path of the index circuit should be calculated, but the whole heating load from each radiator in the system must be added as you work back towards the boiler.

2 An allowance must be made for the resistance created by fittings. This is 30 per cent added to the actual length of pipe to give a figure known as the equivalent length.

3 Where heating pipes run under ground floors or other areas where they do not contribute to the heating of the house, their heat loss must be added to the totals.

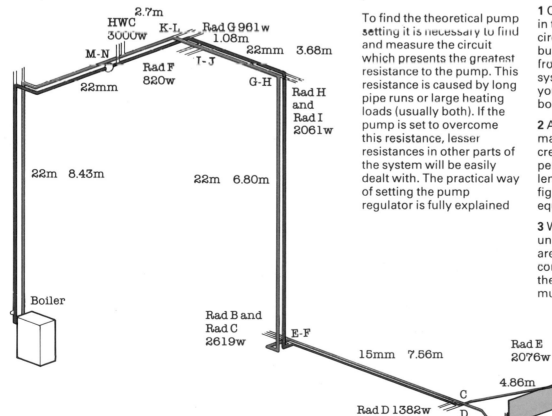

The table (right) gives figures in watts per metre for the different pipe sizes. Where pipes run uninsulated under upstairs floors and through rooms their heat loss can be ignored, since they are helping to heat the house.

Starting at radiator E and using the scale appropriate to the drawing, measure both the flow and return pipes to the tees marked C and D. Write E in the radiator sub-circuit column, AB - CD in the pipe section column, the pipe size (15mm) in the pipe column and the actual length of these two pipes in the actual length column. The length in our example is 4.86m.

Because the pipes run under the ground floor, add the pipe losses in watts per metre, from the table, above right, to the radiator load — 4.86m x 11 (15mm pipe, insulated) equals 53.46 plus radiator E at 2076 watts, equals 2129.46 watts. Round the figure off to 2129 and write this in the column marked 'heating load' and divide by the temperature drop, in this case an invariable 11°C. This gives the equivalent heat flow in watts, 194. Referring to the flow-rate table (right) and the column marked 'equivalent heat flow', you will find the nearest figures are 188-209. You now have to find where 194 falls between these two figures and then read across to the 15mm pipe column and estimate an equivalent figure; 117 would be about right. This gives the pressure loss in Newtons per square metre per metre run of pipe for that section. This is now multiplied by the equivalent pipe length (the actual pipe length plus an allowance of 30 per cent for fittings) 6.32 x 117 = 739 which is the total pressure loss for that section.

Continuing towards the boiler, calculate section CD to EF. For this, add the

heating load from radiator D, 1382 watts to the previous heating total; 2129 plus 1382 equals 3511. 7.56 metres of 15mm pipe plus the pipe loss from the table, 7.56 x 11 equals 83.16 plus the previous heating load makes 3594. Continue as before; 3594 divided by 11 gives 327. From the flow rate table 327 is equivalent to 289 in 15mm pipe. Multiply 289 by the equivalent length 9.83, to make 2840, the pressure loss for the second section.

In pipe section EF to GH, remember to increase to 22mm pipe, so when you refer to the flow-rate table you use the 22mm pipe column. Calculations are as follows: 6.8m pipe length, (no pipe losses are added because the pipes from now on run in heated areas of the house), previous heating load 3594, plus radiators B and C, 820 and 1799, makes 6213. Divide by 11 to make 565, and from the table, 117 is the pressure loss. Multiply this by the equivalent length to make a total circuit pressure loss of 1034.

The next section of pipework is from GH to IJ and, having already carried the heating load from all the downstairs radiators, now add radiators H and I. Previous heat load 6213 plus 679 and 1382 makes 8274 watts; 3.67m pipe length is calculated here. Continuing the sum through the chart, 8274 divided by 11 equals 752, pressure loss from the table - 190 multiplied by the equivalent length gives 908.

IJ to KL is a short section adding only radiator F at 820 watts. Pipe length is 1.08m. Previous heat load, 8274 watts plus 820 makes 9094 divided by 11, 827 and, referring to the table, gives 226. Multiplying by the equivalent length of 1.40 makes 316 — the pressure loss for this section.

KL to MN is a short section at 2.7m, picking up radiator G at 961 watts.

Pipe heat losses

Pipe size (mm)	Exposed (W/m)	Insulated (W/m)
15	43	11
22	63	16
28	77	19

Flow rate table

Pressure loss in N/sq.m per metre run of pipe at 75°C

Flow rate kg/s	Equivalent heat flow watts/°C	Pipe diameter (mm)		
		15	22	28
0.010	42	9	3	
0.016	67	18	4	
0.020	84	27	6	
0.025	105	40	8	2.5
0.030	125	54	11	3.5
0.035	146	71	14	4
0.040	167	90	17	5
0.045	188	110	20	6
0.050	209	132	24	7
0.055	230	155	28	8
0.060	251	181	32	9
0.065	272	209	36	11
0.070	293	237	41	12
0.075	314	267	46	14
0.080	334	299	51	15
0.085	335	335	56	16
0.090	376	370	62	18
0.095	397	406	68	19
0.10	418	445	80	23
0.11	460	527	93	27
0.12	502	616	107	31
0.13	543	1m/s 709	122	35
0.14	585	808	137	40
0.15	627	913	154	45
0.16	669	1.2m/s 1025	171	50
0.17	711	1140	190	55
0.18	752	1263	210	60
0.19	794		230	66
0.20	836		250	72
0.21	878		271	78
0.22	920		294	85
0.23	961		316	91
0.24	1003		340	98
0.25	1045		362	105
0.26	1087		390	113
0.27	1129		417	120
0.28	1170		443	128
0.29	1212		470	135
0.30	1254		500	144
0.31	1296		528	152
0.32	1338			

The index circuit

Radiator sub-circuit	Pipe section	Pipe size (mm)	Actual length (m)	Heating load (w)	Temp. drop (°C)		Equivalent heat flow (w)	Pressure loss (from table) (N/m²m)		Equivalent length (m)		Total pressure loss for section (N/m²)
E*	AB-CD	15	4.86	2129	÷ 11	=	194	117	×	6.32	=	739
E+D*	CD-EF	15	7.56	3594	÷ 11	=	327	289	×	9.83	=	2840
ED+BC	EF-GH	22	6.80	6213	÷ 11	=	565	117	×	8.84	=	1034
EDBC+HI	GH-IJ	22	3.68	8274	÷ 11	=	752	190	×	4.78	=	909
EDBCHI+F	IJ-KL	22	1.08	9094	÷ 11	=	827	226	×	1.40	=	316
EDBCHIF+G	KL-MN	22	2.70	10055	÷ 11	=	914	269	×	3.51	=	944
EDBCHIFG+HWC	MN-BOILER	22	8.43	13055	÷ 11	=	1187	431	×	10.96	=	4724

Total pressure loss for index circuit – N/m²	11506
Divide by 1000 to give pressure loss in kN/m²	11.506

*Includes allowances for pipe heat losses under ground floor

Flow rate: Total heating load divided by the specific heat of water divided by the temperature drop $13055 \div 4200 \div 11 = 0.283$

Pump duty: 0.283kg/s against 11.506kN/m² See pump performance chart

Previous heat load was 9094 plus 961 equals 10055 watts. Divide by 11 to give 914 and, again referring to the table, a pressure loss of 269 is indicated. Multiply by the equivalent length, 3.51, to make 944.

The final circuit is from the cylinder tees MN back to the boiler. Pipe length is 8.43 metres and this section carries the entire heating load so far calculated and the hot-water cylinder loading at 3000 watts. (If your system has a gravity cylinder circuit then you don't include it in the calculations because it is not under the influence of the pump.) 10055 plus 3000 makes 13055, divided by 11 makes 1187. Pressure loss from the table, 431 multiplied by the equivalent length of 11.00 makes 4724.

The totals in the last column are now all added together to give the total resistance of the index circuit in Newtons per square metre. As most pump performance graphs show the pump head in kiloNewtons per square metre it becomes necessary to divide the total by 1000, = 11.505 kN/m².

One other calculation is required to find the pump duty and this is the mass flow rate. This is found simply by taking the total heating load and dividing it by the specific heat of water, which is an invariable 4200, and dividing by the temperature drop of 11°C.

In this example, total heating load 13055 divided by 4200, divided again by 11 makes 0.283. This is the flow rate in kilograms per second. The pump for our system must therefore deliver 0.283 kg/s of water against a total pressure loss of 11.505 kN/m².

The pump performance graph shows the regulator setting for different pump duties. If you read off 11.505 kN/m² from the right-hand side of the graph, and 0.283 kg/s from the top of the graph, the point where they coincide will be the setting for the regulator.

Pump performance graph

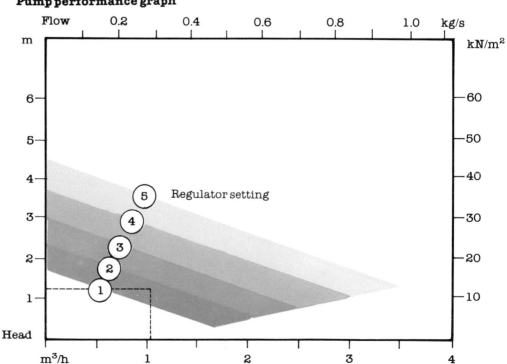

Solid-fuel system

Having worked through the fully pumped gas-fired system on pages 44–5, here is a design exercise featuring a solid-fuel room-heater with the hot-water cylinder on a gravity circuit.

The main problem in a gravity system is running the large-bore pipes in such a way as to give a reasonably quick reheat of the hot-water cylinder. Sometimes you may be lucky enough to find the hot-water cylinder in just the right position relative to the boiler, but as often as not it will be too far away for a neat installation. If the cylinder has to be resited (not so bad if it needs replacing anyway) extra work is entailed, as the cold-water feed and hot-water draw-off pipes will need repositioning in addition to the work involved in the central heating.

When planning, bear in mind the recommendations listed on page 43. It will also be helpful to study the various pipe layouts shown on pages 62–3.

Downstairs layout
The boiler is fitted into an existing fireplace in the living room; this room is not large and the fire will burn continuously, so a radiator is not required.

The long narrow hall benefits from a long low radiator fitted against the party wall, and although one is not shown, the back of the hall could have a small radiator incorporated quite easily if this is considered necessary. This is where familiarity with your own house makes planning easier, as personal experience will tell you where the cold spots and draughts are.

The kitchen has a double radiator fitted near the door. Although the other end of the kitchen under the window would be the ideal place, because of a concrete floor and the shortage of

wall space due to kitchen fitments, a compromise must be made in this small room.

The dining area has a high double radiator fitted next to the French windows to give a high output from a small area. Other positions that could be considered in this room are: on the party wall close to the French windows or along the partition wall next to the hall. It depends mainly on the placing of the furniture and any fixed items such as bookshelves.

Pipe runs are fairly logical on the ground floor. To conceal the pump, a neat cupboard could be constructed in the alcove next to the chimney breast.

Upstairs pipe layout
There are no problems with the upstairs pipe layout, but the hot-water cylinder in its original position in the bathroom would make the gravity pipe run too far away, making it difficult to conceal the 28mm large bore pipes. The solution is to move the hot-water cylinder to the bedroom, as shown. This is convenient for all the pipework; the connections for the upstairs heating circuit can be brought up through here and concealed – the feed and expansion pipe and open vent can be run up through to the roof space and a new airing cupboard built around it all.

An alternative solution would be to have a pumped hot-water system (bearing in mind the special provisions required), leaving the hot-water cylinder in the bathroom but having a 'live' heat leak radiator in the bedroom. (See the layout on the opposite page.)

A useful leaflet describing the necessary controls and pipework for this type of layout is published by Honeywell Control Systems Ltd, and is available from most heating suppliers.

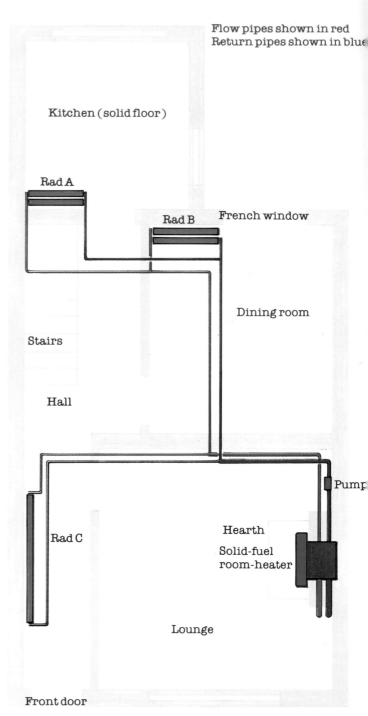

Flow pipes shown in red
Return pipes shown in blue

Kitchen (solid floor)

Rad A

Rad B   French window

Stairs

Dining room

Hall

Pump

Rad C

Hearth

Solid-fuel room-heater

Lounge

Front door

Ground floor

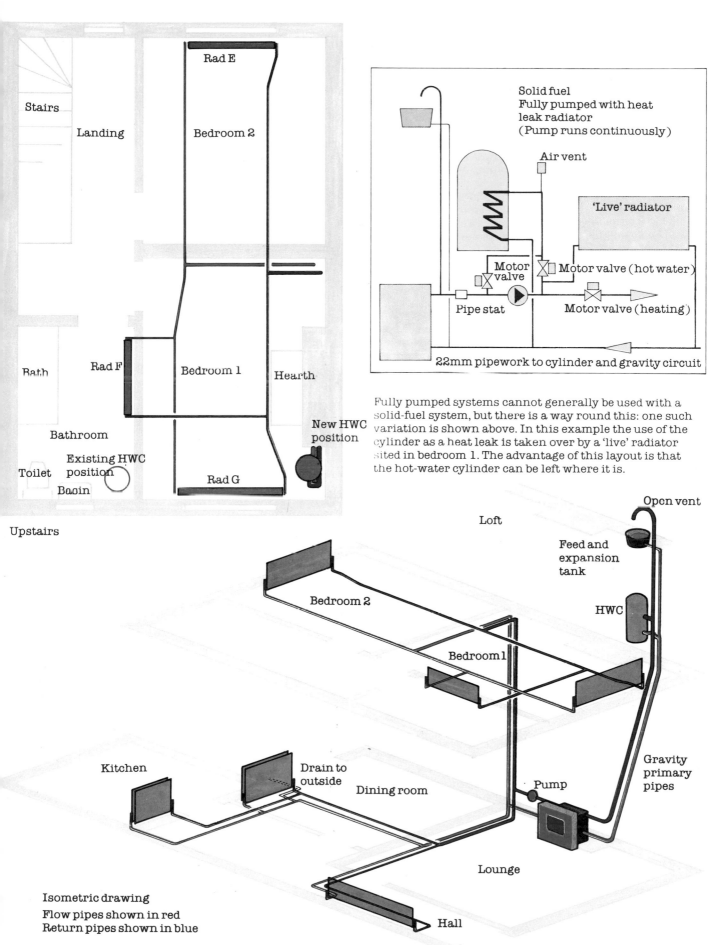

Upstairs

Solid fuel
Fully pumped with heat
leak radiator
(Pump runs continuously)

Air vent

'Live' radiator

Motor valve

Motor valve (hot water)

Pipe stat

Motor valve (heating)

22mm pipework to cylinder and gravity circuit

Fully pumped systems cannot generally be used with a solid-fuel system, but there is a way round this: one such variation is shown above. In this example the use of the cylinder as a heat leak is taken over by a 'live' radiator sited in bedroom 1. The advantage of this layout is that the hot-water cylinder can be left where it is.

Isometric drawing
Flow pipes shown in red
Return pipes shown in blue

Designing for microbore

Before reading this section, read the section on general central-heating design.

Microbore differs from conventional small bore only in the method of distributing the heat to the individual radiators. The basic principle of microbore circuit design for one floor is shown on the right. The heating load is taken to each floor via conventional 22mm or 28mm pipe, the size depending on the heating load of that particular circuit. The manifold is chosen to suit the main pipe size (22mm or 28mm), and the number of outlets to suit the number of radiators on that circuit. All the outlets have 10mm connections and can be reduced to 8mm or 6mm as required. If you think you may be extending your house in the future, fit a ground-floor manifold with spare outlets as these can be very simply blanked off for the present and extra radiators connected up when required.

Installation for a small terrace house

The terrace house shown on page 52, for which we designed a system featuring a solid-fuel room-heater, can also be used as the model for a microbore system. Although a gas boiler is used in this example, microbore systems can be used with all fuels and all types of boilers, gravity primaries or pumped primaries and any size of dwelling; in fact, they are often installed in blocks of flats.

Radiator and boiler sizing is carried out as for a normal small bore installation, so that most of the basic calculations concerning heat losses and so on, considered for the original system, still stand.

Since, instead of a solid-fuel room-heater, a gas room-heater is used, it is necessary to put a radiator

in the lounge because the boiler will not now be operating continuously (as in the case of the solid-fuel boiler) and the room would become cold when unoccupied and the gas fire not in use.

The hot-water circuit

The hot-water cylinder remains as a gravity circuit but the water temperature can now be controlled by a mechanical valve, so when the heating and hot-water circuits are satisfied, the boiler will shut down.

All the pipework can be taken out through one side of the chimney breast as shown in the drawings. This means having large-bore pipework in one recess which will need to be concealed with either a false wall, which can be decorated, or with tongue-and-groove timber cladding.

The ground-floor circuit

The ground-floor circuit shows the manifold placed roughly in the centre of an area formed by the five radiators; the main pipework is therefore run under the floor to here. The plan shows the run to each radiator, but different houses will pose different installation problems that must be taken into consideration when planning, keeping the routes as direct as possible. For example, radiator B shows a slightly longer route for the pipes than if the other end of the radiator were used. It is far easier to remove floorboards from the centre of a room than close to the skirting boards, and it would be difficult to operate the radiator valve in the corner.

The upstairs circuit

The mains pipework for the upstairs heating circuit can rise in the corner of the alcove and run under the upstairs floor to the manifold as shown. Again, the manifold is situated

centrally between the three radiators.

Sizing the pipes

Sizing the pipes that serve the individual radiators requires greater care than for ordinary small bore. This is because the microbore pipe size is a regulating factor in balancing the

system, the aim being to get the resistance as near equal at each radiator as possible. Although there are double-entry microbore valves which include a balancing facility, the more common valve has no balancing control.

First, look at the ground-floor circuit which carries

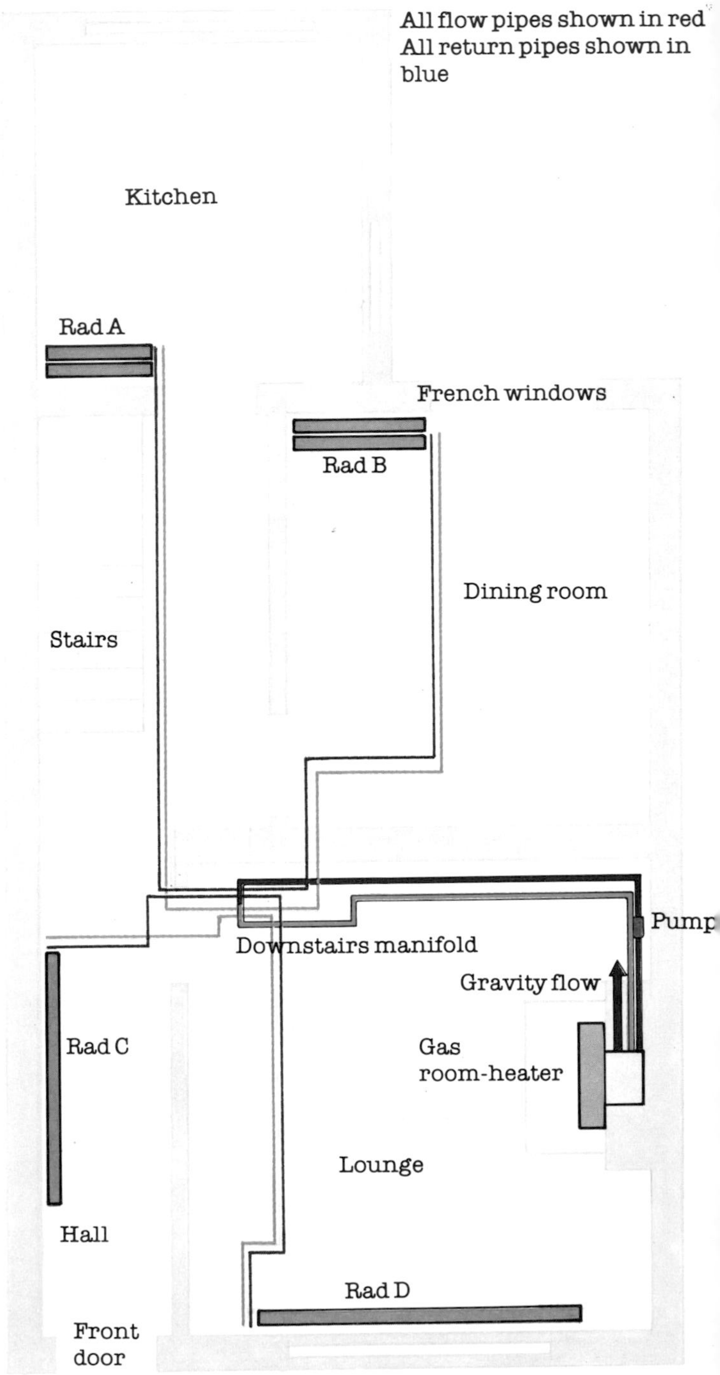

All flow pipes shown in red
All return pipes shown in blue

Kitchen

Rad A

French windows

Rad B

Dining room

Stairs

Downstairs manifold

Pump

Gravity flow

Rad C

Gas room-heater

Hall

Lounge

Front door

Rad D

Ground floor layout

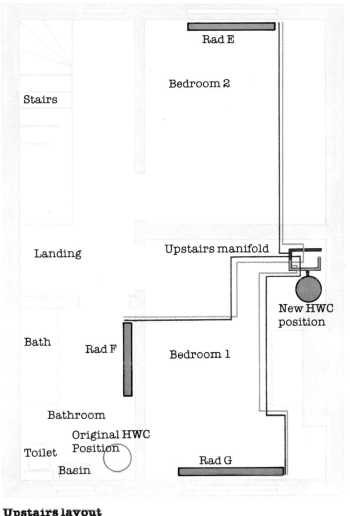

Upstairs layout

four radiators. The pipe size needs to be calculated separately for each radiator, and a simple method of doing this is to take the heating load at the radiator. For example, radiator **A** is 879 watts, and the total length of pipe, both flow and return, is 9.6m; add 30 per cent for fittings. Our requirements are 879 watts and 12.48m, and the table shows 8mm pipe is the size for this sub-circuit. For radiator **B** with a heating load of 1799 watts and a pipe length of 8m plus 30 per cent (10.4m), the table shows 10mm pipe to be the recommended size.

To size the main pipework and select the appropriate size of manifold, add together all the radiator outputs on that circuit and refer to the pipe sizing chart

on page 47. The total load on this circuit adds up to 5555 watts. Although the table shows this load to be just within the carrying capacity of 15mm tube, manifolds are only usually available in 22mm or 28mm sizes and as it's preferable to oversize the mains circuit anyway, to allow the full power of the pump to work on the microbore circuits, a 22mm manifold with eight outlets is selected.

The pump
Certain pump models are specifically designed for microbore installations, as a higher than normal pump head is usually required. Selecting one of these and adjusting it is all that is necessary. However, if the index circuit needs to be calculated to show the

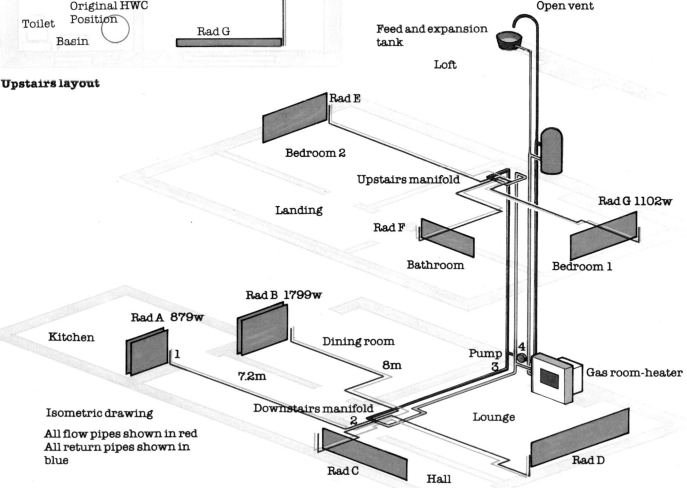

Isometric drawing

All flow pipes shown in red
All return pipes shown in blue

Designing for microbore

pump head, this is the way to go about it.

The index circuit

Studying the section on the small bore index circuit described on page 49 will help with the basic understanding of how this works. It's more difficult with a microbore to spot immediately the index circuit than with a small bore system, so more than one possible circuit may need to be calculated.

Calculating your own index circuit will entail measuring carefully, from your drawing, the length of pipework involved. With our hypothetical system shown here, the measurements are marked on the drawing on the previous page, as well as being shown in the table opposite.

Radiator A is the start of the index circuit on this particular layout, but to give an idea of how this is calculated, read the following explanation.

The actual output from radiator A is 879 watts. To this must be added the heat losses from the length of pipework back to the manifold. This is because the pipework is in an unheated area and does not contribute to the usable heat. Under the pipe sizing calculator on the right is shown the recommended allowance for all microbore pipe of 5 watts per metre. The pipe is 7.2 metres long, so multiplying this by 5 gives 36. Adding this to the radiator load makes 915 watts.

Referring to the details shown at the top of the opposite page and the isometric drawing on the previous page, the radiator sub-circuit is marked A, the pipe section in marked 1–2 and the pipe diameter is shown as 8mm. (This pipe was previously sized under pipe sizing.) The actual length is measured at 7.2 metres (both flow and

return pipes are included) and this is multiplied by 30 per cent, for the resistance of fittings, to give the equivalent length of 9.36 metres.

The heating load, 915 watts, is divided by the temperature drop, 11°C to give the equivalent heat flow, 83. Referring to the flow rate table and looking down the column marked 'equivalent heat flow' until 84 is found, gives a figure as near as possible. Reading across to the 8mm pipe column gives a pressure loss of 903. This is multiplied by the equivalent length of 9.36 to give a pressure loss on this section of the circuit of 8452.

The next section of pipework is the 22mm pipe from the manifold back to the tees near the pump – section 2-3. All the ground-floor radiator loads are added to this section, as well as the heat losses from the pipe, as these carry the whole of that load. (Refer to page 50 for heat losses from pipework other than microbore.) Total load, 6221 watts, divided by 11 gives 566. The pressure loss is found by referring to the flow rate table on page 50 and multiplying this, 115 by the equivalent length, 9.36 shows 1076 to be the pressure loss on this part of the circuit.

The last section is the short run between the tees and the boiler. This carries the entire heating load for the whole house, so this total is calculated as previously to give a pressure loss of 898. These totals are now added together to give a grand total of 10426. Divide this by 1000 to give the total in kiloNewtons per square metre. This figure is on the pump performance graph for your pump.

You can find the mass flow rate by dividing the total heating load by the temperature drop, and

dividing again by the specific heat of water. In this case 8907 divided by 11, divided by 4200 makes a mass flow rate of 0.192 kg/s.

Referring to the pump performance graph and using the pressure loss and the mass flow rate as

calculated will show the pump regulator setting.

Note If one circuit shows a particularly high pressure loss, you may be using too small a pipe size, so using the next size up will give a better figure.

Flow rate table. Microbore pipe

Pressure loss in N/m² per metre run of pipe

Flow rate	Equivalent heat flow	Pipe diameter (mm)		
kg/s	watts/°C	6	8	10
0.007	29	810	182	46
0.008	33	975	210	55
0.009	38	1165	245	68
0.010	42	1395	285	81
0.011	46	1625	325	94
0.012	50	1885	360	107
0.013	54	2165	405	120
0.014	59	2475	460	133
0.015	63	2800	525	145
0.016	67	3155	595	164
0.017	71	3535 (1.5m/s)	665	180
0.018	75	3930	740	195
0.019	79	4350	815	215
0.020	84	4800	903	234
0.021	88	5330	990	253
0.022	92	5780	1075	273
0.023	96	6210 (1.2m/s)	1165	293
0.024	100		1260	313
0.025	105		1350	338
0.031	130		2010	520
0.037	155		2755	754
0.043	180		3585	1030
0.049	205		4485	1325
0.055	230			1670
0.061	255			2030
0.067	280			2410
0.073	305			2830
0.079	330			3280
0.085	355			3780

Pipe sizing calculator. Microbore tube

Boiler flow temperature				82°C	
Temperature drop				11°C	
Maximum flow rate				1.5m/s	
Radiator heating load (w)	Radiator circuit length (m)				
	5	10	15	20	25
500	6	6	6	6	8
1000	6	8	8	8	8
1500	8	8	10	10	10
2000	8	10	10	10	10
2500	10	10	10	10	—
3000	10	10	10	—	—

Microbore pipe heat losses in unheated areas

All microbore pipe – multiply by 5w/m

The index circuit

Radiator sub-circuit	Pipe section	Pipe size (mm)	Actual length (m)	Heating load (w)		Temp. drop (°C)		Equivalent heat flow (w)	Pressure loss from table (N/m²m)		Equivalent length (m)		Total pressure loss for section (N/m²)
*A	1-2	8	7.2	915	÷	11	=	83	903	×	9.36	=	8452
*A + BCDE	2.3	22	7.2	6221	÷	11	=	566	115	×	9.36	=	1076
ABCDE + FGH	3.4	22	3.2	8907	÷	11	=	809	216	×	4.16	=	898

Total pressure loss for index circuit − N/m²	10426
Divide by 1000 to give pressure loss kN/m²	10.43

*Includes allowances for heat losses under ground floor

Flow rate: 0.192kg/s   **Pump duty:** 0.192kg/s against 10.43 kN/m²

Fully pumped microbore system

If you want to leave the hot-water cylinder in the bathroom, where it was originally sited, use a fully pumped system as shown in the diagram below.

The flow and return pipes, incorporating the open vent and the feed and expansion pipes, can rise to the alcove to the first floor. The open vent and the feed and expansion pipe continue up to the loft, while the flow and return run under the floor to junction A where a diverter valve is fitted. Alternatively, two zone valves could be used which will control the heating and the hot water independently.

The main pipework to the downstairs manifold drops behind the short wall in the hall where it can be easily concealed.

Another possible layout, particularly with a solid ground floor, would be to drop the microbore pipes to each radiator from the first floor. In this case a single manifold with 18 connections could be used to serve the whole house, sited as centrally as possible for all the radiators. Drains would need to be fitted to each radiator.

Note Although an ordinary pumped loop is shown here, a cylinder conversion unit fitted to an indirect cylinder could be a suitable alternative.

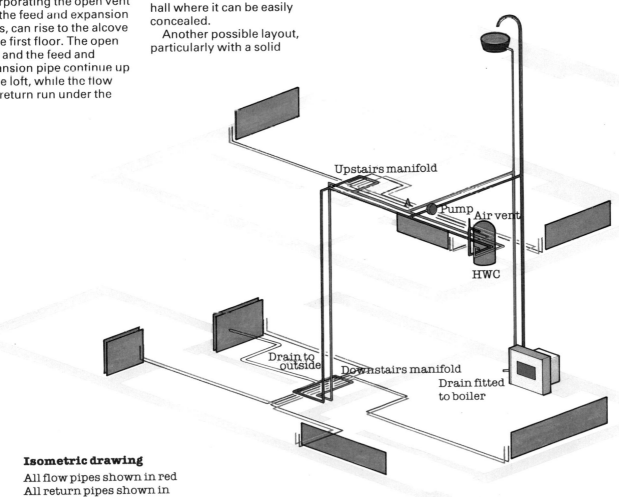

Isometric drawing
All flow pipes shown in red
All return pipes shown in blue

Understanding wiring diagrams

The wiring of controls can sometimes be a difficult part of the installation because, although wiring diagrams are supplied with every electrical control, there may not be sufficient information if your control strategy is anything other than minimal.

The picture becomes much clearer when you purchase your controls as a 'kit' or 'plan' because you then get a data sheet like the one shown on this page which lists every item in the control system and describes exactly which wire is connected to which terminal. The chances of error are much reduced and it will usually cover the whole range of boilers – the wiring for each will be different. These data sheets are constantly updated, so that even the newest boilers are covered.

Make your own large-scale wiring diagram from the information supplied on the data sheet. It can be roughly drawn but must be accurate, so go over it carefully, checking for errors. Use coloured felt-tip pens, and keep the colours to a logical sequence – red for live, blue for neutral and green and yellow for earth. If controls are prewired (as the two zone valves are in this example), follow the colour of any such wiring on your own drawing.

Links, which are sometimes specified between one terminal and another, are simply short pieces of single-strand insulated cable with bare ends, neatly connected between the required terminals. Choose whichever insulation colour suits the circuit.

Although the 1.5mm^2 single-strand twin and earth cable is specified for control wiring, it is clearer to show the colours as already explained, rather than the red, black and bare copper of the actual cable.

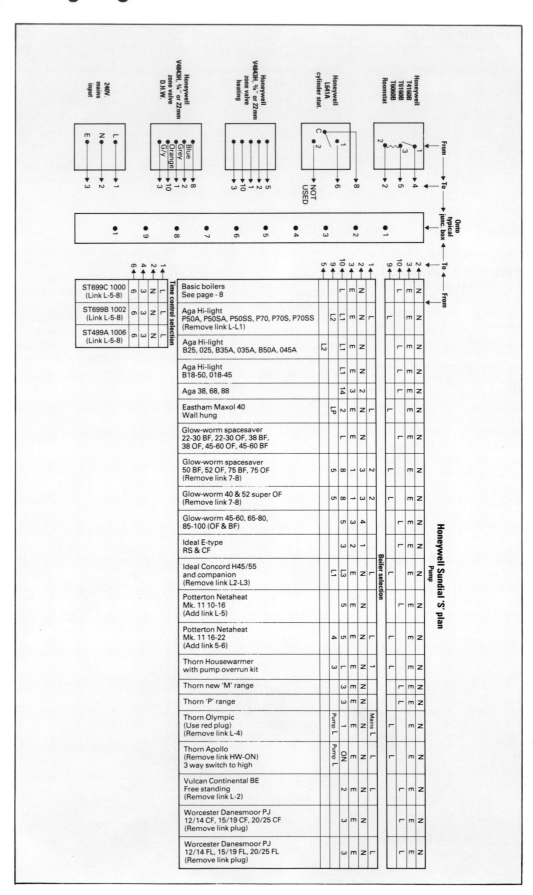

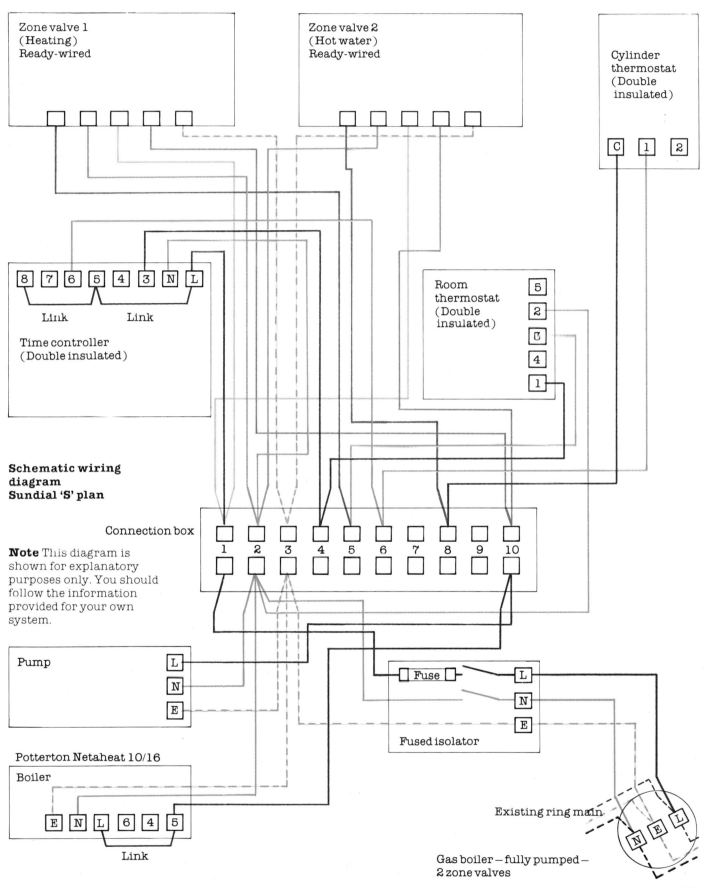

Zone valve 1
(Heating)
Ready-wired

Zone valve 2
(Hot water)
Ready-wired

Cylinder
thermostat
(Double
insulated)

C 1 2

8 7 6 5 4 3 N L

Link   Link

Time controller
(Double insulated)

Room
thermostat
(Double
insulated)

5
2
C
4
1

**Schematic wiring
diagram
Sundial 'S' plan**

Connection box

Note This diagram is
shown for explanatory
purposes only. You should
follow the information
provided for your own
system.

1 2 3 4 5 6 7 8 9 10

Pump

L
N
E

Fuse   L
N
E

Fused isolator

Potterton Netaheat 10/16

Boiler

E N L 6 4 5

Link

Existing ring main

N E L

Gas boiler – fully pumped –
2 zone valves

Designing for pumps and control systems

The siting of the pump requires careful consideration to avoid such problems as pumping over at the open vent or drawing cold water down the feed and expansion pipe.

The ideal pump position is shown in diagram **A1** but individual systems cannot always follow this convenient layout.

The pump can be fitted in either the flow or return pipe; both have their advantages and disadvantages. The return is considered best by some installers since it is usually easier to fit and conceal lower down, and will also run in cooler water. However there is a possibility of air being sucked into the system through minute cracks and leaks, making frequent venting necessary.

Fitting the pump in the flow (although slightly more difficult) means that the system is under positive pressure and any air is pushed to the highest part of the system and vented out.

Generally, pumps can be fitted in any plane so long as the rotor shaft or spindle is horizontal and the pump is not fitted with the motor beneath the pipe where the electrical parts may be susceptible to leaks from the pipework. This means that the pump can be fitted in either horizontal or vertical pipework.

Follow the pump manufacturer's fitting instructions carefully and avoid having the pump in the lowest part of the system where it may collect sediment which accelerates wear.

It is a good idea, when you have chosen your boiler, to obtain a set of installation instructions for the particular model, as it will usually contain useful diagrams of suggested layouts on which you can base your design.

Additionally, particularly in the case of low-water-content, wall-hung boilers, certain mandatory arrangements for pipe sizes and circuit bypasses must be incorporated.

Diagram **A1** shows a basic layout suitable for most boilers of any fuel type except possibly some low-water-content boilers which must be fully pumped. The controls on this, in the case of a solid-fuel boiler, could be simply a time-switch wired to the pump, which would switch off the pump at night when the boiler 'slumbers', any heat going to the hot-water cylinder to be stored and used in the morning. The house temperature during the day would be controlled by the boiler thermostat or individual thermostatic radiator valves. If the boiler was fired by gas or oil, more controls could be fitted.

The house temperature could be controlled by a room thermostat and programmer and the hot-water temperature by a mechanical cylinder control valve.

Diagram **A3** shows a similar layout with the cylinder circuit controlled by a motorized valve. An alteration to the pipework is necessary to allow unrestricted flow between the feed and expansion tank and the boiler, a major safety factor.

Diagram **C** shows a fully pumped system which allows greater flexibility and ease of installation due to the use of smaller bore pipework. This is a fairly straightforward, reasonably economical system to install and run. It will give good control over both space and water heating. The basis of this system is a two-way diverter valve which will supply heat either to the heating circuit or the hot-water circuit according to demand. The cylinder circuit usually has priority which

means that in periods of high demand for hot water the heating takes second place. A bypass may be required (see the boiler manufacturer's instructions). An interesting feature of this design is that no hot-water allowance needs to be added to the heat requirements when sizing the boiler, as the full output is available to heat the water at the expense of the heating circuit.

Also available is a three-way diverter valve which supplies heat to either the heating circuit, the hot-water circuit, or both. In this case, the hot-water allowance must be included when sizing the boiler.

Fully pumped systems are not generally applicable to solid-fuel installations, but there is a system whereby the usual purpose of the cylinder as a heat leak has been directed to a radiator. This allows the cylinder to be fully pumped and temperature controlled as with gas or oil. One such layout is shown in **diagram B**, in which the pump runs continuously while the

boiler is running; 22mm pipes are required for the cylinder circuit, the bypass and the 'live' radiator circuit, and special provisions for the feed and expansion tank to withstand possible high temperatures.

Diagram **D** shows a fully pumped system with separate motorized zone valves to control the flow to either the heating or hot-water circuits. Each zone valve is controlled by its own thermostat through the programmer.

Diagram **E** shows a sophisticated system controlled by three zone-valves. One valve controls the hot water while the other two control individually the downstairs heating and the upstairs heating. This means that the upstairs heating can be programmed to come on later than the downstairs heating just before the family go to bed. A simple override button on the programmer can switch on upstairs heating if, for example, children go to bed early or wish to work in their bedrooms.

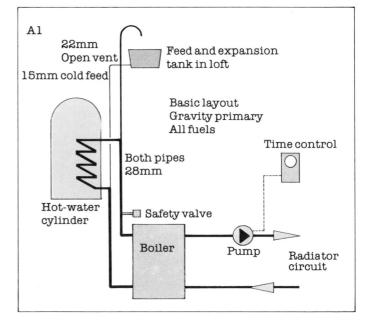

Above: A basic layout suitable for most boilers and fuel types.

Right: A selection of layouts suitable for different applications.

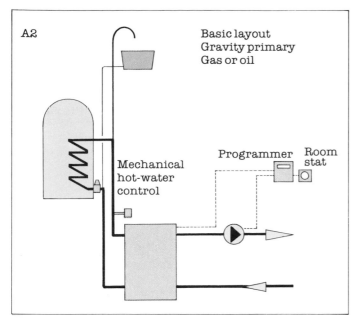

A2 Basic layout
Gravity primary
Gas or oil

Programmer Room stat

Mechanical hot-water control

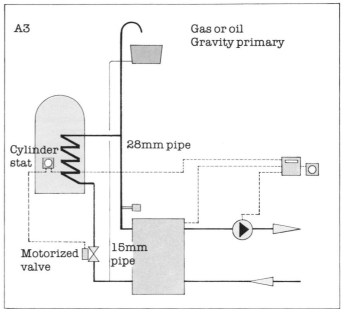

A3 Gas or oil
Gravity primary

Cylinder stat

28mm pipe

Motorized valve 15mm pipe

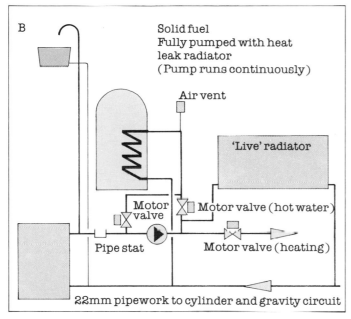

B Solid fuel
Fully pumped with heat leak radiator
(Pump runs continuously)

Air vent

'Live' radiator

Motor valve Motor valve (hot water)

Pipe stat Motor valve (heating)

22mm pipework to cylinder and gravity circuit

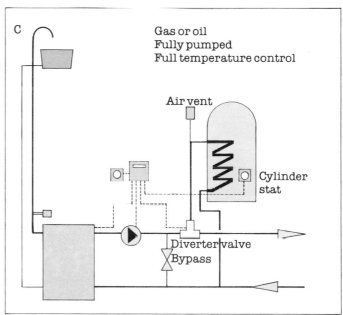

C Gas or oil
Fully pumped
Full temperature control

Air vent

Cylinder stat

Diverter valve
Bypass

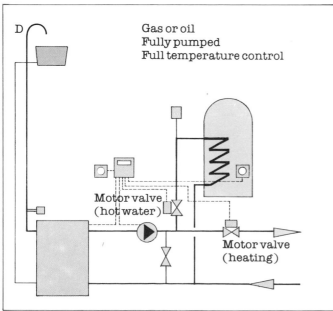

D Gas or oil
Fully pumped
Full temperature control

Motor valve (hot water)

Motor valve (heating)

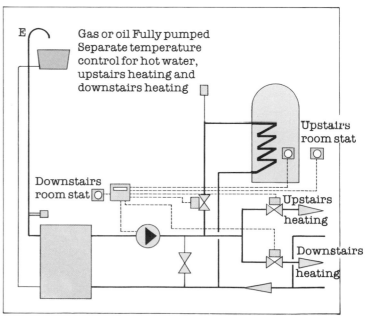

E Gas or oil Fully pumped
Separate temperature control for hot water, upstairs heating and downstairs heating

Upstairs room stat

Downstairs room stat

Upstairs heating

Downstairs heating

Basic design layouts

You can adapt and extend these layouts when designing your own system.

From the information already given, you should have no difficulty in designing a suitable layout for your own dwelling. If you have special problems not covered in this book it would be advisable to consult a heating designer.

For example, if you have no loft a sealed system could be considered. Instead of the usual feed and expansion tank this uses a small sealed container with a flexible diaphragm to take up the expansion and contraction of the heating water. As special equipment is required for this type of installation, it requires professional help.

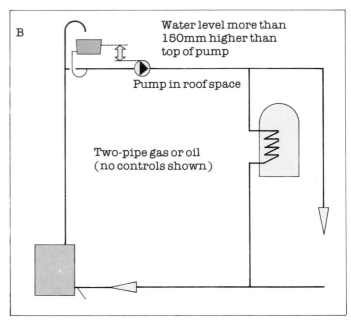

A Two-pipe system for gas or oil (no controls shown)

Wall-hung boiler

Pump in return

A Two-pipe system for a gas wall-hung boiler with gravity hot water and the pump in the return.

B Water level more than 150mm higher than top of pump

Pump in roof space

Two-pipe gas or oil (no controls shown)

B Two-pipe fully pumped gas- or oil-fired system with the pump in the loft. The feed and expansion pipe must be looped under, and the minimum water-height level observed.

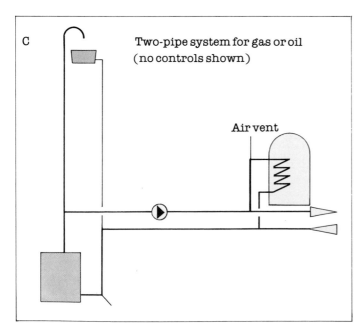

C Two-pipe system for gas or oil (no controls shown)

Air vent

C Two-pipe fully pumped system suitable for gas or oil boilers. An air vent is required at the cylinder and a drain at the boiler. The two circuits could be controlled by a two-way or three-way diverter valve, or separate zone valves.

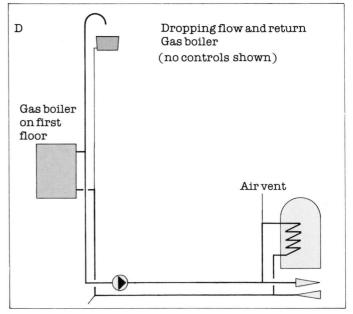

D Dropping flow and return Gas boiler (no controls shown)

Gas boiler on first floor

Air vent

D This gas boiler is situated on the first floor, with the pump on the ground floor: a fully pumped system with suitable controls as in 'C'.

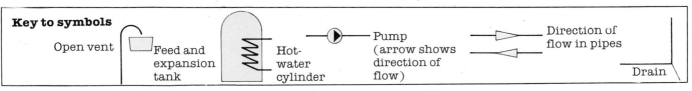

Key to symbols

Open vent   Feed and expansion tank   Hot-water cylinder   Pump (arrow shows direction of flow)   Direction of flow in pipes   Drain

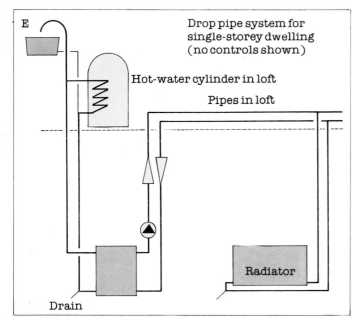

E   Drop pipe system for single-storey dwelling (no controls shown)

Hot-water cylinder in loft

Pipes in loft

Radiator

Drain

E A system suitable for a bungalow with solid floors. With a gravity hot-water system this would be suitable for all fuel types. Air vents are required on high points of the pipework and drains on all the radiators.

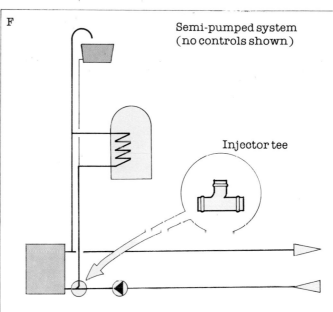

F   Semi-pumped system (no controls shown)

Injector tee

F Semi-pumped gravity hot-water system using special injector tees to assist the gravity flow while providing pumped heating. It is suitable for all fuel types. The pump is fitted in the return.

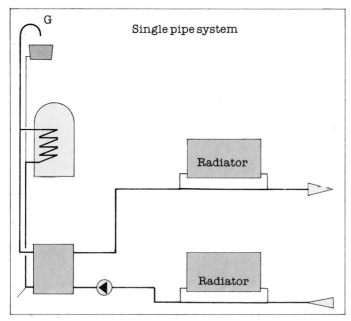

G   Single pipe system

Radiator

Radiator

G Single-pipe system. This old type of system is not recommended but is included here so that a home-owner who has one will recognize it.

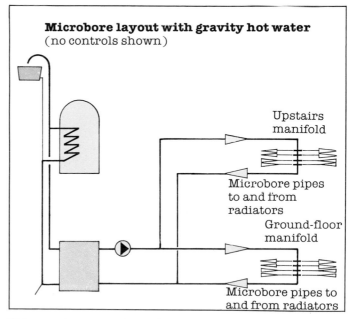

Microbore layout with gravity hot water (no controls shown)

Upstairs manifold

Microbore pipes to and from radiators

Ground-floor manifold

Microbore pipes to and from radiators

H A microbore system showing a manifold serving each floor. This layout shows a gravity hot-water system, but with suitable adjustments to the pipework, could be made a fully pumped system.

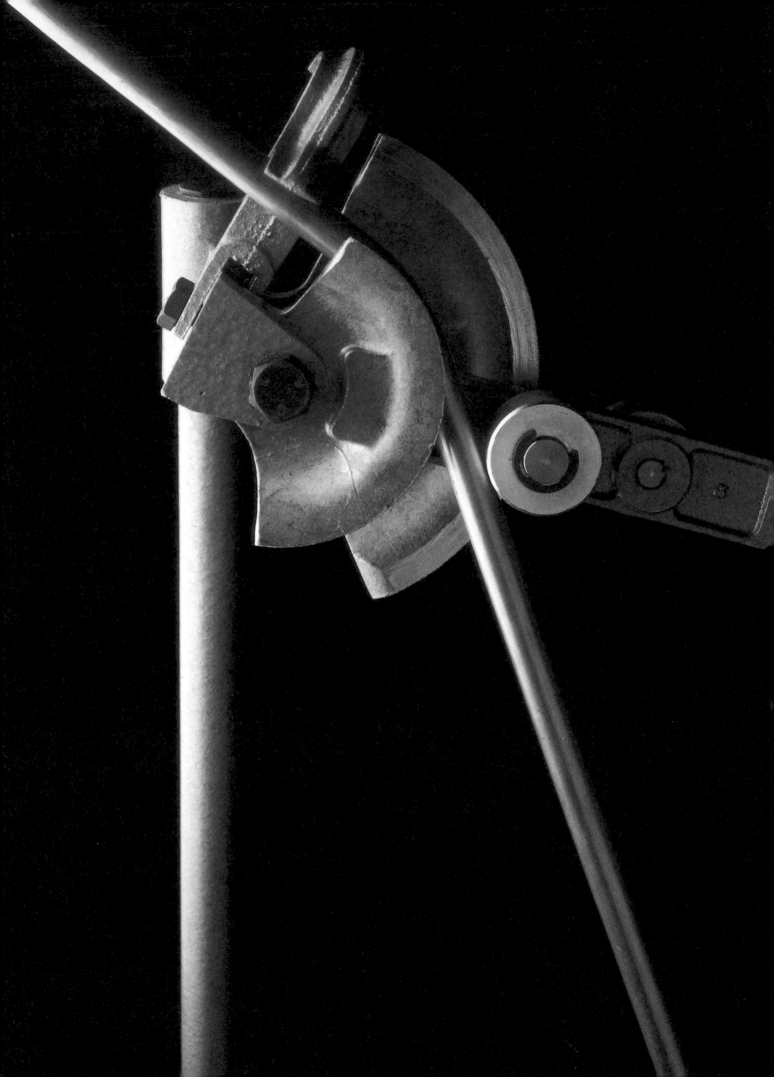

INSTALLING CENTRAL HEATING

Tools to buy

It cannot be overemphasized that the right tool makes any job easier. Although buying tools is expensive, most of those listed here will be useful additions to the home tool chest.

1 Pipe cutter: Although copper pipe can be easily cut with a hacksaw, it is almost impossible to cut the end square. This tool does that job perfectly, and also has a reamer at the end to remove burrs from the cut.

2 Hexagon wrench: Necessary for fitting some kinds of radiator valves.

3 Adjustable spanner: For tightening threaded fittings.

4 Hacksaw: For cutting old pipe and pipe in positions where a pipe cutter cannot be manipulated.

5 Water pump pliers: For general use and threaded fittings.

6 Pipe wrench (Stillsons): For tightening threaded fittings, particularly on boilers.

7 Pipe cleaner: Removes swarf from inside pipe ends.

8 Files: 250mm (10in) flat, and round for cleaning and preparing pipe ends.

9 Pipe-bending springs: 15 and 22mm for bending pipe.

10 Glass fibre mat: For protecting joists etc from scorching when using the blowlamp; you could use a large, ceramic tile for the same purpose.

11 Gas blowlamp: This is more precise and controllable than the old paraffin type.

12 Steel measuring tape: Get a locking type.

13 Bold marker pen: For marking pipes and floorboards for cutting.

14 PTFE tape: Very thin, conformable tape for sealing threaded joints.

15 Solder: Sold in 1/2 kg reels specifically for copper pipe.

16 Flux: For use with solder.

17 Wire wool: For cleaning pipe.

18 Jointing compound: For use on compression fittings.

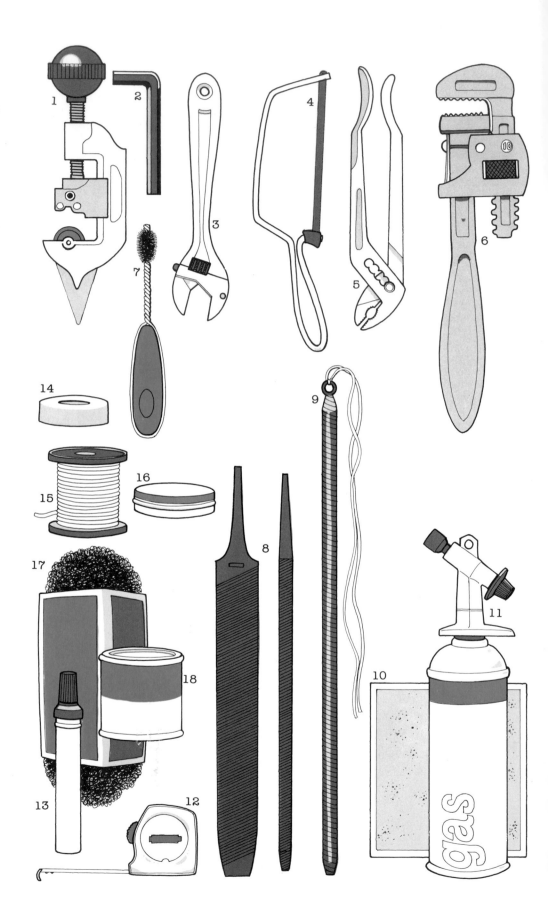

Tools to hire

Some specialized tools can make certain jobs in a central-heating installation a great deal easier, but are probably not worth buying. They can be hired, either from hire shops or from large central-heating discount warehouses.

1 Pipe-bending machine: Essential for bending 28mm pipes (if used), and also useful for smaller sizes.

2 Powerful slow-speed drill (and large long-masonry bit): This greatly simplifies drilling holes through brick walls.

3 Immersion-heater spanner: For removing old immersion heaters from hot-water cylinders and fitting new ones.

Bending pipe

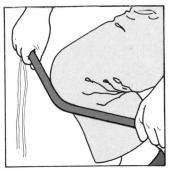

1 Mark the approximate centre of the bend on the pipe. Aim to get the centre of the spring at the centre of the bend.

2 If the spring has to go right in, tie a cord to the eye and drop it into the pipe.

There are two ways of changing the direction of a pipe run: one is to use fittings and the other is to bend the pipe itself. Bending is preferable, particularly with the smallest pipe size, since there is less friction, less work and less chance of a leak. Copper pipe is very malleable and the 15mm is very easy to bend with a bending spring. The 22mm is slightly more difficult, and the 28mm really requires the use of a bending machine. As well as the large, floor-standing bending machines, shown above, smaller hand-held machines are also available from hire shops. Even with 15mm pipe, however, it is unwise to make bends of small radius because, besides splitting the pipe, you are likely to damage the bending spring. In this case use elbow fittings instead. If a bending spring will not go easily into a pipe, check to see if the pipe has been damaged. If it has, cut off the offending part.

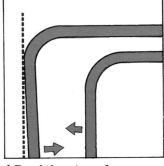

3 Hold the pipe as shown and pull it towards you; check the angle frequently.

4 Bend the pipe a few degrees beyond a right angle and ease it back. This releases the spring.

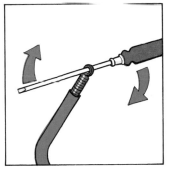

5 If the spring will not come out, put a screwdriver in the eye and twist it clockwise.

When measuring and marking the pipe, always allow extra at the ends — it is almost impossible to measure exactly where the bend will be, and it is better to lose a few inches of pipe than to finish with a uselessly bent one that is too short.

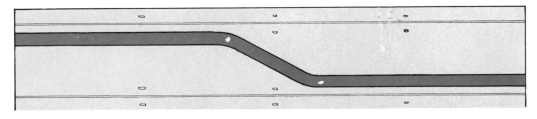

6 To keep a pipe run parallel while putting a bend in it to avoid an obstruction, first mark the width of the bend on a floorboard and make the first bend. Check against the mark and make the second bend, then check for parallel against the floorboard.

Fitting radiators

Preparation

You save yourself great difficulty later if you paint the bottom and backs of the radiators before installation. Although radiators come finished in a light coat of white primer, this is not sufficient protection in steamy locations such as bathrooms or kitchens. They will soon rust if not properly painted: a good quality undercoat followed by a gloss finish will protect them from corrosion.

A certain amount of heat from the radiator passes slowly through the wall; this loss can be reduced by fitting special foil to the wall before the radiator is hung so that heat is reflected back into the room. It is obtainable from DIY merchants and heating warehouses.

Since radiators are heavy when full of water, it is important to get a good fixing when fitting the brackets to the wall. In some old houses the plaster is very soft, so drill well into the underlying brick to ensure that the wall-plug bites into something solid.

Hanging radiators

Do not forget to add 100mm to the measurement between the bottom of the radiator and the bracket, to allow for clearance between the floor and the radiator.

When positioning brackets on the vertical lines marked on the wall, do not align the bracket holes with the lines. It is the angle of the bracket (the part which sticks out from the wall) which must be in line with the vertical.

The brackets supplied with the radiators come with one round hole and one slotted hole to allow for adjustment up and down. So, when fitting the second radiator bracket to the wall, mark only through the centre of the slotted hole. Fixing the bracket temporarily with a single

screw allows you to adjust it to get the radiator level. Once it is correct, drill and plug the second hole and fix the radiator permanently. Try to arrange it so that the pipes from the valves down through the floor will not foul the joists. This is not always possible, and a slight bend in the pipe may be necessary.

Where special problems occur due to the shape of a room, refer to page 96.

The vent is usually supplied fitted to the radiator, but if it is not, the blanking plug at the highest end of the radiator should be removed and the vent screwed in after treating the threads with PTFE tape.

The valves can be fitted either before or after hanging the radiators. If you do it afterwards, the radiator is held securely on its brackets and the tail of the valve can be tightened easily. Should the position of the radiator make fitting awkward, lift it off its brackets and fit the valves.

When reassembling the valve, leave it temporarily finger tight — it is much easier to measure pipes accurately for cutting and also easier to drill the hole in the floorboard if the valve can be removed. The valve can be permanently fixed when the pipes are run.

Refer to your plan and ascertain which end will be 'flow' and which 'return'. Take the wheel-valve and unscrew the large nut which secures the threaded male end to the radiator body, and separate.

Treat the threaded end with PTFE tape and tighten it securely (with the nut attached) into the radiator tapping at the flow end. The method of tightening these parts varies with different makes of valve. Reassemble the valve and leave finger tight only. Treat the lockshield valve similarly.

1 Lay the radiator, face down, with the bottom against the skirting board. Align the edge of a builder's level along the centre of both welded supports and mark the wall. Where it is impossible to lay the radiator flat, lean it out from the wall, sight down, and mark the bracket positions.

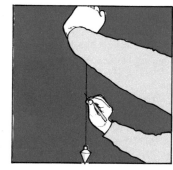

2 Draw a line at each point using the builder's level to ensure accurate verticals.

3 Alternatively use a plumb line; mark the wall and use a straight edge to draw the lines.

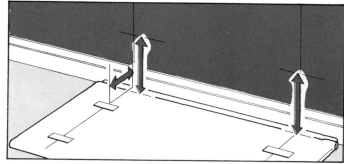

4 Measure from the bottom of the radiator to the bottom of the lower welded support. Add 100mm and mark the wall up from the floor to the vertical lines.

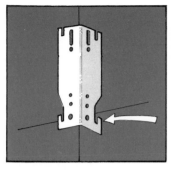

5 The bracket is lined up with the vertical line, and the bottom of the lower hook with the mark.

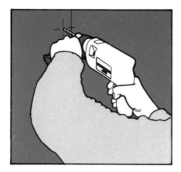

6 Mark through the centre of the holes and drill the wall with a masonry bit.

Fitting the valves

1 Separate the tail of the valve from the body by unscrewing the large nut.

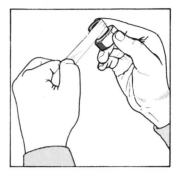

2 Treat the thread of the tail with PTFE tape and insert it into the radiator.

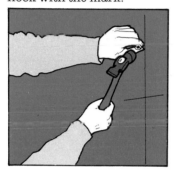

7 Hammer in a suitable wall plug and screw the bracket tightly to the wall.

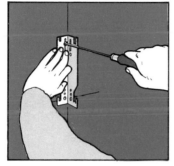

8 Fit the second bracket: a single screw in the slotted hole will allow for adjustment.

3 Tighten it into the radiator with a hexagon wrench.

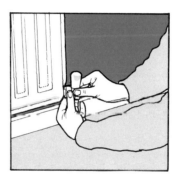

4 Reassemble the valve to the tail and leave it finger tight only.

9 Place the level on top of the radiator and, by adjusting the second bracket, allow the radiator to rise very slightly at the vent end. When satisfied, mark the second hole and drill and fix.

Tools and materials

Builder's level
Plumb line and straight-edge
Soft pencil
Power drill (the hammer type is best for a lot of masonry drilling)
Masonry drill bit; wall plugs and screws of correct size for radiator brackets
Screwdriver; steel

measuring tape; hammer
PTFE tape

Depending on the type of valve used, either:
   a 15mm hexagon wrench
   or
   a spanner
   or
   pliers and wrench

Radiator valve fittings

A In this type, instead of the usual circular bore, the interior of the tail is machined to a hexagon. A 15mm hexagon wrench is required to tighten it into the radiator.

B This one has four flats machined on the exterior and is simply tightened by a spanner. Do not use a pipe wrench or water-pump pliers – they could damage the chrome finish.

C This type has a round bore with two key ways. The proper tool is a union wrench, but you could improvise by selecting a tool with an end that fits tightly in the bore and using a wrench to turn it.

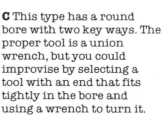

Joining pipe with capillary fittings

Take time and care to make soldered joints properly. A few extra minutes spent now will save hours later, if you have to drain the system and dry out pipes to cure a leaking joint.

Clean inside the ends of pipes and fittings and tap out the filings. Fine pieces of metal left in the system when it is filled can affect the sealing of drain plugs and valves as well as damaging pump bearings. A smooth interior of pipes and fittings also reduces friction and makes for a more efficient system.

When cutting pipe do not apply excessive pressure, since this will cause the pipe to deform and also makes a heavy burr.

It is much easier to solder pipe fittings in a vice, so wherever possible prefabricate as much pipework as you can, leaving the final connections to be made on site. Always check, however, that you can fit a piece of made-up pipework into position first.

End-feed fittings

End-feed fittings are very similar to other capillary types, the difference being that end-feed fittings have no integral solder ring, so you have to apply solder to the heated and fluxed parts.

The pipe ends should be prepared as described for capillary fittings, but should be tinned: apply solder to the heated pipe end and brush or wipe it off while it is still running. The pipe should have a complete silvery coating. Apply more flux and push the pipe end into the fitting, heat it and then add solder to the junction of the pipe and the fitting.

Tools and materials

Pipe cutter; black waterproof marker; files; wire wool; flux and solder; gas blowlamp; cloth; fibreglass mat.

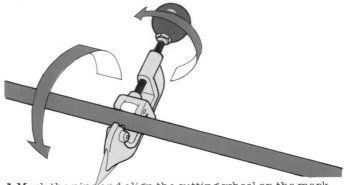

1 Mark the pipe and align the cutting wheel on the mark. Adjust so that the cutter lightly grips the pipe. Swing the cutter around the pipe, gradually tightening down the knob to cut through it.

2 Remove the burr from the pipe end with the blade of the cutting tool.

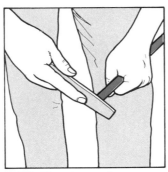

3 File the pipe end square with a large flat file.

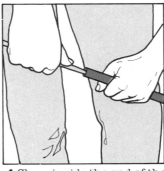

4 Clean inside the end of the pipe with a smooth round file.

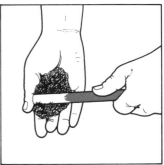

5 Clean the end of the pipe thoroughly with wire wool. Do not handle the end.

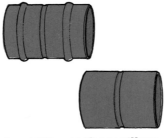

In addition to the capillary fitting, which has an integral ring of solder at each end, there is the end-feed fitting.

6 Clean the inside of the fitting thoroughly with wire wool.

7 Apply flux liberally to the inside of the fitting and the outside of the pipe.

8 Heat up the fitting with the pipe in place until an unbroken ring of solder appears.

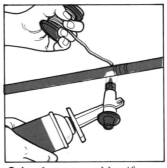

9 Apply more solder if necessary by touching it on the hot pipe.

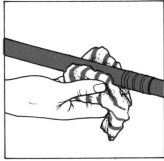

10 Allow to cool thoroughly and wipe off the excess flux to avoid corrosion.

Joining pipe with compression fittings

A typical compression coupling. Prepare the pipe as described for capillary fittings opposite. Do not apply flux.

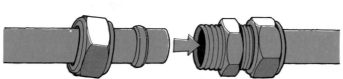

1 Slide one nut on to the prepared pipe end, open thread towards the pipe end. Push the olive on to the pipe about 25mm from the end.

Pipe joining with compression fittings requires only spanners or wrenches – no blowlamp or solder.

Compression fittings are bulkier than capillary ones, and more expensive, but they can be separated at any time. Also, leaking joints usually require only tightening, not the draining down of the system as in the case of a leaking soldered fitting.

If you decide to use compression fittings, there will still be parts that require soldered joints. The average system will contain both types of joint – and you will have to choose which type suits the particular situation.

2 Coat the olive and the pipe end thinly with jointing compound.

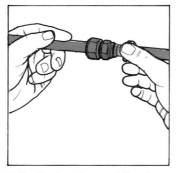

3 Push the pipe end firmly into the fitting until the pipe meets the internal shoulder, and screw down the nut.

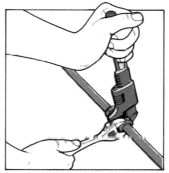

4 Hold the centre nut with a wrench and tighten the end nut firmly with an adjustable spanner.

Tools and materials

Pipe cutter; black waterproof marker; files; wire wool; adjustable spanners; jointing compound and PTFE tape.

Threaded fittings

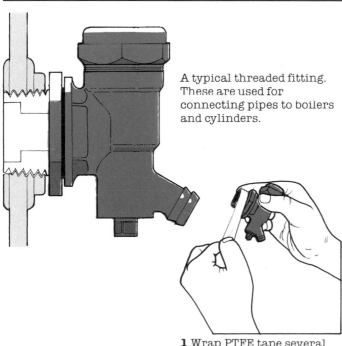

A typical threaded fitting. These are used for connecting pipes to boilers and cylinders.

1 Wrap PTFE tape several times around the thread, finishing with the end of the tape pointing away from the direction in which the fitting is screwed.

2 Tighten the fitting with an adjustable spanner or wrench.

Threaded fittings are used where pipework is to be connected to boilers and cylinders. The threaded part is usually tapered so the more the fitting is screwed in, the tighter it becomes.

PTFE tape is used on the threads for a seal. Start the tape on the thread with the end facing towards the direction into which the fitting is to be screwed and wind it around several times. On large threaded joints, be very liberal with the PTFE tape – it is very thin and threads are not always machined to very fine tolerances.

Clean out old jointing compound, tape and hemp from threads etc., when connecting to existing items.

Installing the boiler/solid-fuel room-heater

These instructions are to give the reader a general idea of the type and amount of work entailed in fitting a solid-fuel room-heater. Full and detailed instructions are supplied with all boilers and they must be followed carefully. The work shown here is a condensed version of the instructions for fitting a Parkray 111GL room-heater.

Provided that the existing fire surround is to your taste and has a reasonably flat surface immediately around the fire opening, it will be satisfactory for the installation of a room-heater.

As the holes in the chimney breast cannot be made good until the system is tested for leaks, fitting the boiler in this case could be made the final part of the installation. The holes would not then need to be left open for a long period.

The chimney must be in good condition and should be swept before starting work. Solid-fuel room-heaters do not necessarily require a liner for the chimney, but gas room-heaters do.

Observe the building regulations concerning hearths and non-combustible materials.

When the whole central-heating installation is complete and tested for leaks, and the chimney breast repaired, a small fire can be lit and the radiators bled and balanced.

Vermiculite concrete
This is required for an insulated infill around the boiler and flue. It is made with vermiculite granules, Micafil or similar (as used for loft insulation), available from DIY merchants.

Recommended mix
Six parts of vermiculite to one part of cement should be used. Mix thoroughly together and add only

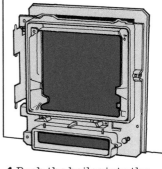

1 According to your pipe layout, one or both sides of the chimney breast will need access holes. Measure the height of the pipe openings in the boiler and remove only sufficient bricks to allow access. The front hole is to enable infilling around the fire and the fitting of the flue spigot. Firebacks can be broken up and removed without damage to the surround. Make sure the recessed hearth is level.

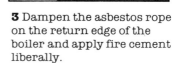

2 According to your requirements, blank off any unused boiler tappings, treating the plugs with PTFE tape first.

3 Dampen the asbestos rope on the return edge of the boiler and apply fire cement liberally.

4 Push the boiler into the opening, hard up against the fire surround.

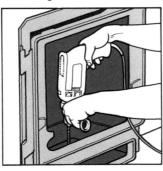

5 Drill through the base casting with a 6.5mm bit. Drill into the hearth with a No. 12 masonry bit.

6 Push the plug into the hole and screw in the anchor screw. Cover with fire cement.

7 Make the boiler connections through the side access hole, using PTFE tape on the male thread.

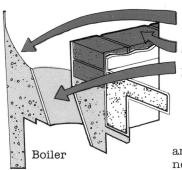

Vermiculite backfill
smoothed into chimney

Access hole

Flue forming pipe

Boiler

8 The complete system must be tested for leaks before infilling can begin. Cover the front of the boiler,

and push a ball of newspaper into the flue opening. Shovel vermiculite concrete into the area around the boiler. Tamp it down with a stick. Fit a

short length of 150mm diameter flue pipe and smooth the concrete into the chimney.

sufficient water to hold the mixture together when squeezed in the hand.

When infilling round the boiler back and sides, tamp down the mortar with a stick to avoid air pockets.

When the cavity round the boiler is filled, the flue pipe should be fitted and the area around this filled too. Then smooth the vermiculite concrete to form a seal from the top of the flue pipe into the chimney proper.

The opening in the front of the chimney breast can now be bricked up and made good.

9 Brick up the opening and allow it to dry. Make sure there are no cracks.

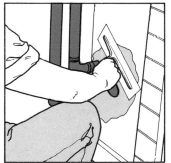

10 Spread a rendering coat of mortar just below the plaster surface and allow it to dry.

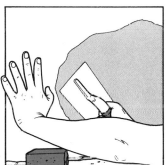

11 Use a straight edge to level the plaster and finish by smoothing with a dampened steel float.

Tools and materials
Club hammer
Bolster chisel
6.5mm drill bit
Builder's bucket for removing rubble
Dust sheet
Wrenches
PTFE tape
Fire cement
Hammer drill and No. 12 masonry bit
Screwdriver
Pointing trowel and vermiculite granules
Cement
Steel trowel.

12 On no account must the pipes fall away from the boiler on the gravity cylinder circuit.

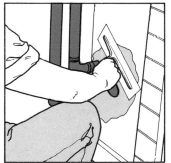

13 Fill the remaining cavity with brick rubble and brick up the opening. Finish as before.

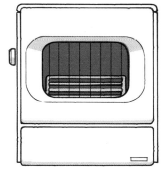

14 Reassemble all the fire parts and remove the newspaper from the flue spigot.

Removing an old fireplace surround

Most fireplace surrounds are attached to the wall by lugs at the side. After removing the screws, lever the surround away from the wall with a wrecking bar; take care, it will be very heavy. It may require breaking up before it can be removed from the room. The hearth can usually be separated from the sub-hearth with a shovel.

Installing the boiler/wall-hung gas boiler

These instructions are to give the reader a general idea of the type and amount of work entailed in fitting a gas boiler. Full and detailed instructions are supplied with all boilers and these must be followed carefully. The work shown here is a shortened version of the instructions for fitting a Potterton Netaheat 10/16 wall-hung, balanced-flue gas boiler.

Wall-hung gas boilers can be fitted at any height; but if fitted in a kitchen, it is best to align the base with the base of the wall units. If wall units are not yet fitted, the recommended height is 450mm above the worktop. Worktops are usually 900mm high.

When attaching the template to the wall, use either masking tape or panel pins. A really good fixing is essential for a wall-hung boiler — use expanding bolts if necessary.

Cutting a hole through a solid brick wall is heavy work, made more difficult by the fact that chisels are generally shorter than the wall thickness — you need to work from both sides. The problem is to find the centre from inside and out. The best way is to measure from the edge of an adjacent window or door, and strike a vertical plumb line. With a long straight-edge and level, mark an accurate horizontal. Measure up or down from where these two lines cross and make a small, tentative hole through the centre of the cut-out.

Another way is to use a powerful slow-speed drill and long masonry bit. Drill a series of holes and chop out in between.

Obstructions may be encountered in the wall when cutting the flue hole, so do this job first. If the boiler position needs to be adjusted to one side or the other, the mounting holes can be drilled to suit.

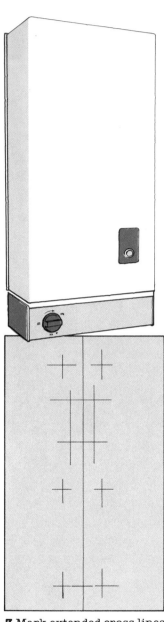

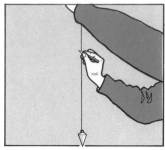

1 The boiler and its equipment comes in two cartons. Two metal feet (discarded later) are bolted to the boiler to protect the controls. Unpack the cartons and identify the parts as described in the boiler instructions. Read the instructions carefully before starting work. A cardboard template is supplied for marking the holes and the flue opening on the wall. Having chosen the boiler position and ensured that it conforms to the recommendations concerning flue outlets, mark a centre plumb line on the wall.

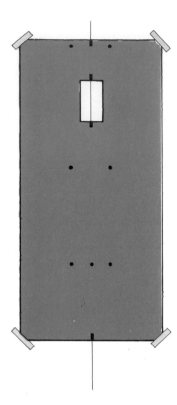

2 Attach the template securely to the wall with the centre marks on the vertical line.

3 Mark extended cross lines at the hole centres and the wall cut-out.

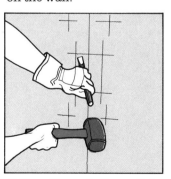

4 Chop through the wall for the flue and check the liner is an easy sliding fit.

5 Drill all the holes well into the brickwork and fit the plugs.

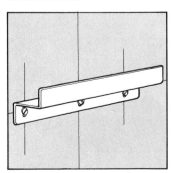

6 Screw the mounting channel tightly to the wall and check for level.

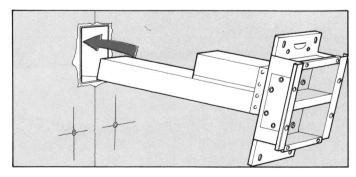

7 Bolt the duct assembly temporarily to the plenum chamber, with the shortest duct uppermost, and push the whole assembly tight up against the inside wall.

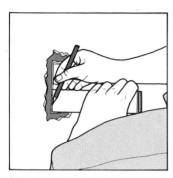

8 From outside the house, mark where the longest duct protrudes from the wall and add 41mm.

9 Mark all round the duct with a try square and saw off the excess with a hacksaw.

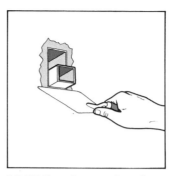

10 Fit the plenum chamber and ducting and make good the walls, inside and out.

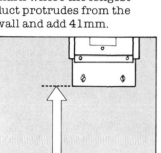

11 Measure the distance as shown and add or remove shims as required.

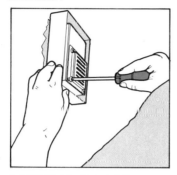

12 Fit the balanced-flue terminal, sealing it with the strip supplied.

13 Secure short lengths of pipe to the back of the boiler, according to the layout.

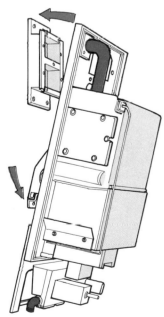

14 Check that all the work done so far is satisfactory and complete. This boiler requires two people to lift it on to its support. Lift and position the boiler centrally on to the mounting channel, push the top against the plenum chamber and secure with the screws supplied, noting that some of the screw holes are for the fan cover. Reassemble the fan, wiring the flue hood as described in the instructions for your particular boiler. Remove the metal feet.

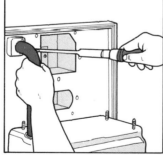

14 Check that all the work done so far is satisfactory and complete. This boiler requires two people to lift it on to its support. Lift and position the boiler centrally on to the mounting channel, push the top against the plenum chamber and secure with the screws supplied, noting that some of the screw holes are for the fan cover. Reassemble the fan, wiring the flue hood as described in the instructions for your particular boiler. Remove the metal feet.

15 Dismantle the gas cock, and run a length of pipe near the supply pipe.

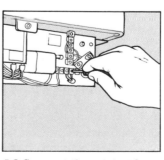

16 Connect the wiring for the chosen control system, and run it in PVC conduit wherever it is exposed.

When cutting through a cavity wall, try to avoid knocking pieces of brick into the cavity, as this can create a bridge between the two walls which, in some circumstances, can lead to damp penetrating the inner wall.

If the wall is lightweight block, cutting through it will be easy as this material can even be sawn.

Do all the necessary making good to the internal wall before assembling the boiler in position.

Safety precautions
Take precautions when cutting through walls. It is tiring for the wrist and if your concentration goes you are likely to hit your hand instead of the chisel. A thick glove worn on that hand will soften the blow. Wear safety goggles to guard your eyes.

Gas connections
It is unlawful for unauthorized persons to tamper with the gas supply, so the final connection must be done by an approved gas fitter. This can be done at the time of commissioning the boiler, and as this requires special equipment, the whole work can be entrusted to an approved fitter.

Tools and materials
Plumb line or builder's level
Pencil; masking tape
Cold chisels, wide and narrow
Club hammer
Gloves and goggles
Hammer drill and masonry bit
Screwdrivers
Measuring tape
Try square
Hacksaw
Trowel and rendering mix
Filling knife and filler
Wrenches
PTFE tape
Blowlamp
Flux and solder
Wire wool
Hammer

Installing a flue-liner

Flexible, stainless-steel flue-liners should be used only for gas or oil installations and never for solid fuel. If a chimney needs to be lined for a solid-fuel installation, use an approved type.

Although installing a flue-liner is a relatively simple job, the prospect of working high on the roof is not always appealing. If you have any doubts employ a builder.

Preparing the chimney

If a chimney pot is fitted to the proposed flue outlet it must be removed. Take care – pots are much larger than they seem from the ground.

The next job is to check the chimney for obstructions. Lower a weighted paint can of a slightly larger diameter than the liner down the chimney. If you encounter an obstruction, it will usually be where the chimney changes direction. By measuring the rope to the obstruction you should be able to locate roughly the position inside the house. Knock out a brick or two from the chimney breast at this point and clear the flue.

Measuring

Stainless-steel flue-liners are available cut to length, or in kits. Drop a weighted string down your chimney and add 300mm for every bend. Be generous with your estimate. Liners come in diameters of 100mm, 125mm and 150mm. Check your boiler literature for the correct size.

Installation

With a rope attached to the nose-cone of the liner, gently feed it down the chimney while an assistant pulls the rope from below. Don't pull too hard or the liner may be damaged. If it is necessary to cut the liner to size, wrap newspaper around the end as a guide to sawing it square, and use a hacksaw.

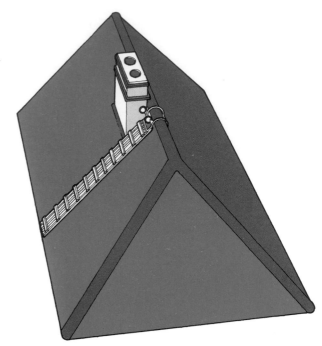

When working on a ladder at an exposed chimney, lash the ladder securely to the top. If you hire a scaffold tower you will be secure.

Use a proper roof ladder, which hooks over the top of the ridge, when it is necessary to climb over a roof. Use a scaffolding tower to climb on to the roof ladder and not an ordinary ladder.

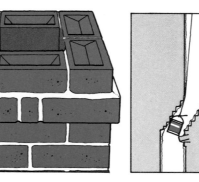

1 Clear the top of the chimney roughly to the level of the brickwork.

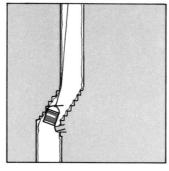

2 Check for obstructions by lowering a weighted can of a slightly larger diameter than the liner.

3 Feed the liner down the chimney with a rope attached to the nose cone.

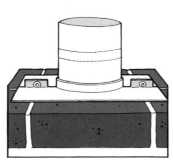

4 Allow the liner to protrude 150mm and clamp the locking plate in position.

5 Cover the clamp plate with a rendering mix and smooth off at the edges. Fit the terminal.

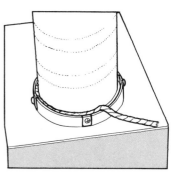

6 Fit the liner into the flue spigot and seal with asbestos rope and fire cement.

Oil storage tank installation

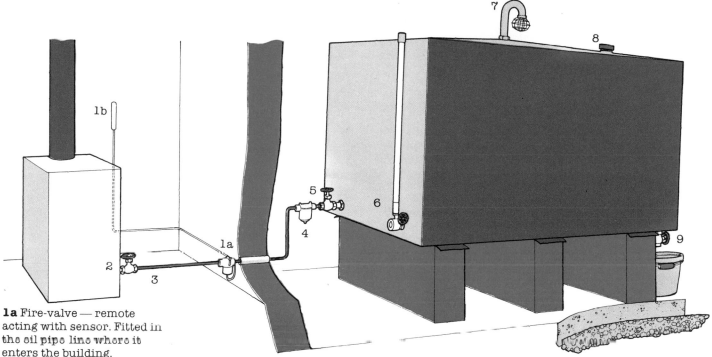

1a Fire-valve — remote acting with sensor. Fitted in the oil pipe line where it enters the building.
1b Fire-valve sensor. Fitted near the boiler, 1.5m from the floor.
2 Isolating valve. Shuts off the supply to the boiler.

3 Oil pipe-line (10mm copper).
4 Oil filter (automatically filters the oil supply).
5 Shut-off valve. Shuts off the supply from the tank for filter removal.
6 Sight gauge. Indicates the oil level in the tank.
7 Air vent (for equalizing

the pressure in the tank).
8 Filling pipe. 50mm male thread hose coupling with a non-ferrous screw cap.
9 Drain valve.

Additional work and expense is involved in the installation of an oil-fired system because the fuel needs to be stored in a large volume. The tank should be positioned so that the take-off for the oil pipe-line is between 380mm and 760mm higher than the burner in the boiler. (Boiler manufacturers' instructions give specific details.)

A 25mm fall away from the take-off end of the tank should be allowed for when building the piers. The brick piers should be not more than 1.5m apart and should stand on a concrete base 150mm thick on a bed of well-tamped hardcore, also 150mm thick, extending slightly beyond the concrete. A damp-proof membrane should be inserted between the brickwork and the tank.

Tanker access
The delivery tanker needs to

be able to reach the oil tank with its filling hose. If the tank is too far from the road or drive (will your drive support a 20-tonne tanker?), an offset pipe can be made up in 38mm pipe with 50mm male screw-connector and cap.

Minimum distance from boiler
The tank should be sited as close to the boiler as possible (bearing in mind any local by-laws) and preferably no more than 15m away.

Local planning consent
Once you have decided where the tank is to be sited, make a simple drawing showing distances, sizes and thickness of concrete bases. This must be approved by the local planning officer. Some councils require a catch-pit capable of containing the entire contents of the tank in

the event of a leak, but you will discover what the local authority's requirements are only if you contact the planning office directly.

The tank
Oil storage tanks are available in various shapes and sizes – 1100 litres (250 gallons) to 3400 litres (750 gallons) direct from large central-heating suppliers. A charge is usually made for delivery.

The oil line
This is usually made from 10mm soft copper as used for microbore heating systems. For pipe runs longer than 15m, use 15mm pipe. The pipe should be supported at 1.5m intervals with pipe clips. If the pipe is buried it should be run through PVC pipes.

Tank connection kit
An oil-tank connection kit is a very good idea as it will

contain all the necessary ancillary equipment for a tank, complete with all the connectors and nipples as well as 7.5m of oil-feed pipe.

Fire-valve
The type with a remote sensor is the safest, as the sensor can be sited close to the boiler to react quickly.

Sludge valve
The tank should slope slightly away from the feed pipe end, and a valve should be fitted to the lowest part to drain off sludge and water. Allow room for a bucket.

Sight gauge
A sight gauge shows the current contents of the tank so a dipstick is not needed.

Open vent
A vent pipe fitted with a wire balloon is situated at the highest point of the tank to allow displaced air to escape during filling.

Removing floorboards

Tools

1 Club hammer (any fairly heavy hammer will suffice but this tool will be useful for other parts of the installation, such as knocking holes through walls).

2 Bolster chisel (a wide, strong chisel, ideal for levering up floorboards).

3 Tenon saw (suitable for most flooring jobs, but a floorboard saw, which has a blade with a rounded end, is even better).

4 Trimming knife (fitted with a heavy-duty blade, it will cut through the tongues of tongue and groove floorboards).

5 Wood chisels (for chopping out the slots in joists for the pipes. You will need 12mm and 19mm chisels).

6 Padsaw (a useful alternative to the trimming knife for cutting floorboard tongues).

7 Electric jigsaw (not essential, but necessary for method C of floorboard removal).

8 Circular saw (for method B of floorboard removal – fit a general-purpose blade).

It may be possible to punch nails right through a floorboard and into a joist to make removal easier. If you have fitted carpets it may be necessary to call in a carpet fitter to take up the carpet and refit it in due course. If you have sanded and varnished floorboards it may be possible, to preserve the finish of the floor around the edge of the room by running a prefabricated pipe from under an area covered by rugs or furniture, and bringing it up through a hole beside the radiator.

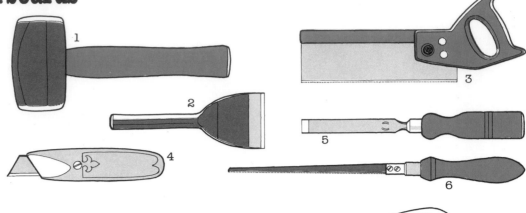

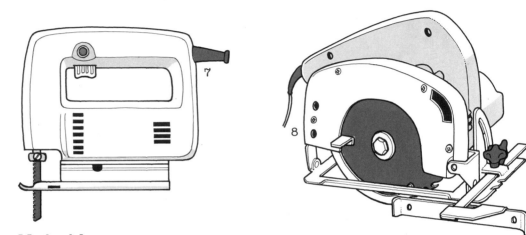

Method A

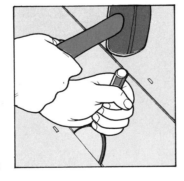

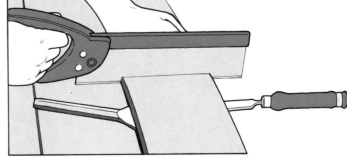

Hammer in chisels at either side of the joist to lever up the floorboard and wedge it. Saw across with a tenon saw and do the same at the next selected joist.

Method B

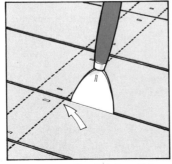

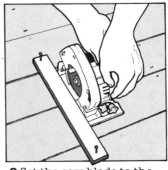

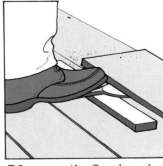

1 To locate the joists, slide a scraper blade between the cracks and mark the edge of the joist on the top face of the boards.

2 Set the saw blade to the exact depth of the floorboard. Temporarily nail a batten as a guide and saw across the floorboard.

3 Lever up the floorboard with chisels.

Method C

Hold the jigsaw firmly and start in the position shown, beside the joist, not above it. Allow the saw to cut down through the floorboard. Do the same at the next selected joist. With this method it is necessary to screw battens to the side of the joist to support the shortened floorboard.

Warning — this method should be used only where you are certain there are no pipes or cables running over the top of that particular joist.

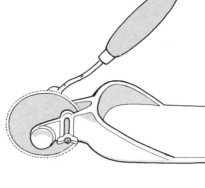

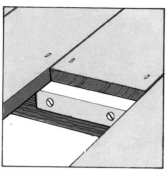

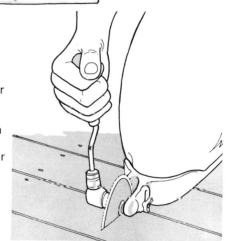

Floorboard cutter

This is a very useful, specialized tool available for hire from some heating-supplies warehouses. It uses a circular saw blade on a ratchet handle with a baseplate, on which the user can exert his weight while working the handle, to cut through the floorboard. A very safe tool which neatly solves the problem of floorboard removal.

Running pipes beneath floors

Various types of plastic pipe-clips are available in the three common sizes: 28mm, 22mm and 15mm.

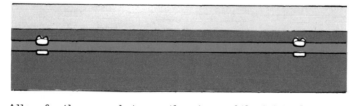

Allow for the space between the pipe and the joist when connecting the pipe. It will then slip neatly into the clip without force.

Pipes in floor spaces should be clipped neatly to the sides of joists where they run parallel to them, and dropped into cut-outs where they run across them. Room must be allowed for expansion, so never wedge a pipe angle into a confined space. Leave room for lateral expansion, especially on long straight runs.

To avoid weakening the timber when cutting slots in joists for pipes, try to avoid running pipes through notched joists across the centre of a room — keep to the side. For the same reason, do not saw too much out of them but make them only just large enough for the pipe and insulation. The method shown here is suitable for pipes in upstairs rooms (which don't need to be insulated). In ground-floor installations the pipes can run under the joists.

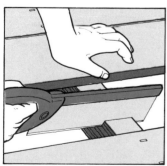

1 Mark the position and depth of the pipe with a marker pen. Do not saw below the depth mark.

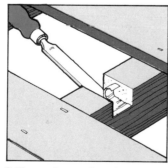

2 Use a sharp chisel to remove the waste between the saw cuts, and check with a pipe offcut for depth.

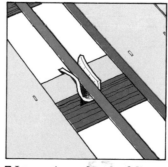

3 Lay a piece of underfelt or plastic foam under the pipe for insulation against noise.

Running the pipes

It is best to start at the radiator furthest from the boiler and work back towards it, picking up connecting pipes as you go.

Remove all the furniture from the room you are working in and take up any floor covering. Refer to your floor plan and decide where the pipes will run and take up floorboards where necessary. Remove all protruding nails from joists and floorboards before starting work.

If you have to leave any unjoined pipes to work somewhere else, mark these pipes with an arrow, using a bold waterproof marker pen, to show the 'flow' and 'return'.

Make up sections of pipe without soldering initially — you can then adjust the length or bend of each pipe before finishing them. Try to prefabricate as much pipework as possible — it is much easier to solder a joint in a vice than under the floorboards.

It is almost impossible to solder one end of a capillary joint without the solder at the other end running: a tee, for example, will need all its pipes prepared and in position before you apply any heat. If you must solder a joint with one end open, soak a heavy cloth in water and wrap it around the unused part (alternatively, use an end-feed or compression fitting).

When connecting pipes through floorboards to a radiator, make sure the vertical pipe comes right up the shoulder in the valve connection. Use jointing compound for the pipe to the valve connection. Keep all your short ends of 15mm pipe — they are useful for radiator connections.

When pipes run alongside joists they should be clipped high on the side of the joist, so that when they change direction and go across it is not necessary to notch the joist too deeply.

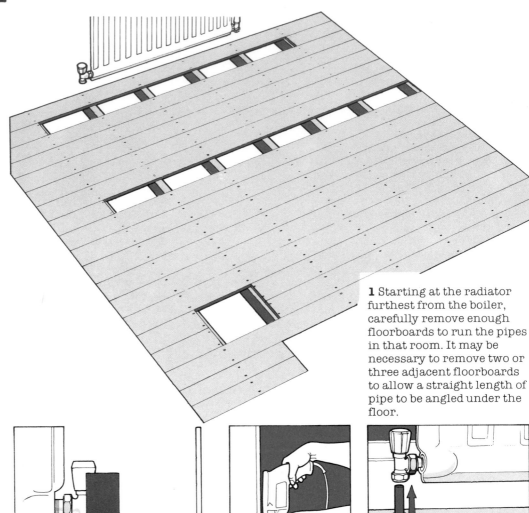

1 Starting at the radiator furthest from the boiler, carefully remove enough floorboards to run the pipes in that room. It may be necessary to remove two or three adjacent floorboards to allow a straight length of pipe to be angled under the floor.

2 Align a set square with the centre of the valve pipe entry and mark the floorboard.

3 Remove the valve body and drill through the floorboard using a flat bit.

4 Reassemble the valve temporarily. Make up an unsoldered section of pipe and cut the vertical pipe exactly to size.

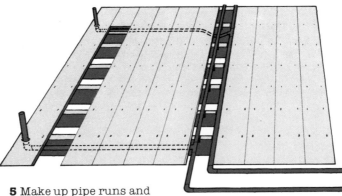

5 Make up pipe runs and assemble them before soldering any joints.

6 When you have to use a blowlamp close to joists, soak the surrounding woodwork with water first.

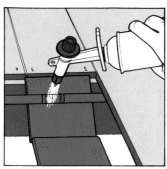

7 Use old ceramic tiles or a fibreglass mat to protect against the scorching of joists and floorboards.

8 When a joint is made, check the back of it with a torch and mirror. Add more solder if necessary.

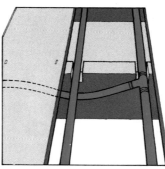

9 To kink a pipe to run it under another, use the bending spring.

10 Under suspended ground floors, clipping pipes underneath the joists saves sawing slots in them.

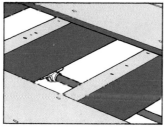

11 Use a large clip and wrap with insulation before pinning to the underside of a joist.

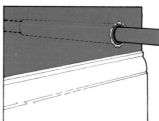

12 When running pipes through walls above floors, sleeve the pipe through a length of larger bore pipe.

When running pipes under suspended ground floors it is not necessary to notch the joists, as the pipes can be attached to their undersides. Pipes can also be run more directly at an angle.

If you have to push a pipe through a sleeper wall under the floor, tape up the leading end with masking tape, since a piece of debris in the pipe could cause endless problems.

Fire safety
Take precautions when using a blowlamp beneath floorboards. Use an old washing-up liquid bottle to squirt water on the joists and floorboards around where you are making a joint.

Position ceramic tiles, a roof tile or a fibreglass mat to prevent flames playing on to bare wood. Keep the water bottle within easy reach when using a blowlamp.

1 Mark the pipe run on the floor and chop out a channel to the required depth.

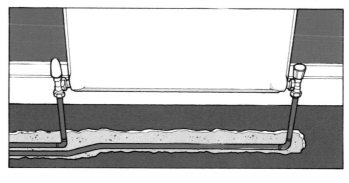

2 Check by laying a straight length of pipe in the channel and chop out any obstructions. Leave the pipes uncovered until the system is tested for leaks.

Solid floors
Solid floors present a special problem. Although floor screeds are relatively soft it is worth avoiding making long runs across them. In some cases it may be possible to attach pipes neatly to skirting boards or drop them from the floor above to a particular radiator, running them behind curtains if the radiator is under a window, although both these methods use considerably more pipe and will alter the dynamics of the system. If your house has solid floors throughout, a drop-feed system using microbore pipe could be the answer. See pages 54-7 and 92-3.

Pipes buried in concrete must be protected against corrosion, so cover them with Densotape before filling the channel with mortar.

3 If possible, wrap the pipe with Densotape (or similar) before connecting.

4 Alternatively, before filling the channel, cover the pipes with Densotape.

5 Fill the channel with a mix of mortar and level it with a steel float.

The hot-water cylinder

It will be helpful to read the section on domestic hot water in the design section.

If you have an indirect cylinder (the type with an inner coil heating the domestic hot water), you should not need to touch the domestic hot-water system at all. Just connect up the primary pipes from the boiler to the cylinder according to your plan.

If you have a direct cylinder (the type in which the directly heated water is drawn off by the taps) and it is in good condition, you may be able to adapt it to your new heating system. Using a heating element fitted into the immersion heater opening, a direct cylinder can be easily converted to an indirect cylinder.

If your existing cylinder is at all suspect and shows signs of leaking (greenish-blue deposits around joints), it makes sense to replace it.

The connections on old cylinders are sometimes difficult to remove. Heating them with a blowlamp before applying pressure with a wrench will help. The walls of copper cylinders are thin and liable to distort if extreme force is used.

A recent development in cylinder design is an insulated coating applied after manufacture: cut away the insulation for contact if a cylinder thermostat is to be fitted. These are highly recommended, as it is almost impossible to insulate a normal cylinder as effectively as those with a polystyrene coating. One problem arises, however: if female secondary connections are involved, particularly the cold feed to the bottom of the cylinder, there will be insufficient room around the thick insulations to manipulate a spanner or wrench, so some sort of extended connection is required. Enquire about this when you buy your cylinder.

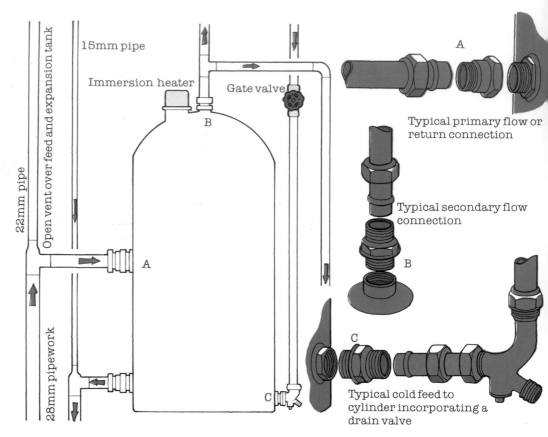

Pipe layout for gravity hot-water system

Typical primary flow or return connection

Typical secondary flow connection

Typical cold feed to cylinder incorporating a drain valve

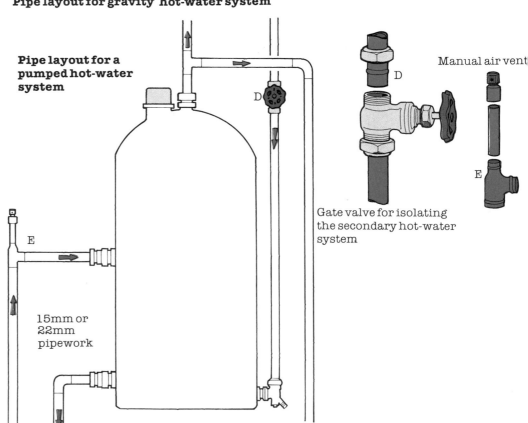

Pipe layout for a pumped hot-water system

Manual air vent

Gate valve for isolating the secondary hot-water system

Replacing a hot-water cylinder

1 Close down the gate valve controlling the cold-water feed to the cylinder.

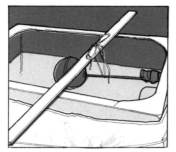

2 If a valve is not fitted, tie up the ballcock in the cold-water storage tank.

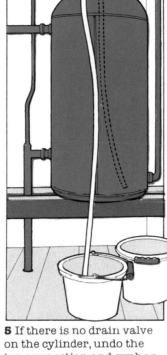

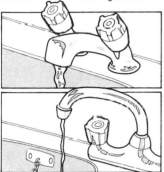

3 Open all the hot taps, starting at the highest and finishing with the lowest.

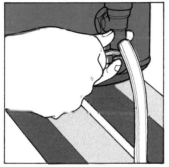

4 Fit a length of hosepipe to the outlet and open the drain valve.

5 If there is no drain valve on the cylinder, undo the top connection and syphon out the water.

6 When the cylinder is drained, undo all the connections.

7 If the old connections will fit, clean off all old jointing compound before reconnecting.

8 If an immersion heater is not being fitted, cap the boss with a blanking plug.

9 If an immersion heater is being used, fit it using the special spanner.

10 To clear an air lock, follow the instructions on the right.

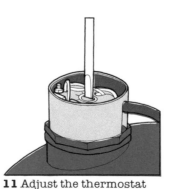

11 Adjust the thermostat on the immersion heater to the required setting.

Immersion heaters

If your heating system is run by gas or oil it is probably best to heat your domestic hot water in the summer from the boiler. If you have a solid-fuel boiler however, it is more convenient to use an immersion heater in the summer. Even if you are using gas or oil, it is a good idea to fit an immersion heater, as it will give you an independent source of supply if the boiler is ever out of order.

Immersion heaters have built-in thermostats which should be set to the required water temperature according to the instructions for that particular heater. The usual setting for domestic hot water is 60°C.

Clearing an air lock

Should the hot water not run freely after work has been completed, the likely cause is air trapped in the pipes. Remove the cover from the cold-water tank so that the vent pipe is free to eject water into the tank. Select a cold tap that is fed directly from the mains and connect up a short piece of hosepipe from it to a hot tap. Secure the hose tightly to both taps. Close all the other taps and open the connected hot tap fully.

Open the cold tap fully. Get an assistant to watch the vent pipe in the loft and signal when the water is forced fiercely out of the vent pipe. Let it flow for a few seconds then turn off the cold tap. Turn off the hot tap and remove the hose. Check that all the hot taps flow freely.

The feed and expansion tank

For clarity, the tank and all the pipes are shown uninsulated. All pipework and tanks in loft areas should be insulated.

The feed and expansion tank should be situated in the loft as nearly as possible over the boiler, to keep the pipe runs short. It should be installed at least 1200mm above the top of the highest radiator or heating pipe. Normally this will present no problem, but if siting the tank on the loft floor does not satisfy this requirement, construct a strong, simple platform to gain height.

A 45-litre feed and expansion tank will be suitable for most systems up to about 22kw. Prepare all the fittings to the tank before taking it into the loft. Make sure all holes are cut according to your pipework positions. Do not make these holes too large: it is virtually impossible to repair a hole in a polythene tank. Some tanks are supplied with a metal plate on the outside of the fitting to counteract distortion of the tank by the ball valve. In this case the position of the inlet will be dictated by the plate.

When the tank is ready, rig up a light in the loft and clear the area where you will be working. Polythene tanks require a solid base to stand on, so make a suitable-sized platform from 19mm blockboard or plywood, and nail it in position on the joists.

The feed and expansion pipe runs from the tank to the boiler, either directly or via the hot-water cylinder according to your system. (The section on hot-water cylinders shows this more clearly.)

The inlet is connected into the rising-main water supply to the cold-water tank — usually this is convenient. Do not connect into the cold-feed supply pipe to the hot-water cylinder because this is not at mains pressure. Ball

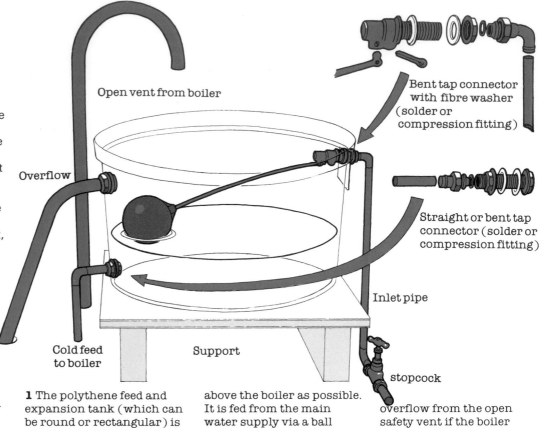

Open vent from boiler

Overflow

Bent tap connector with fibre washer (solder or compression fitting)

Straight or bent tap connector (solder or compression fitting)

Inlet pipe

Cold feed to boiler

Support

stopcock

1 The polythene feed and expansion tank (which can be round or rectangular) is sited in the loft as nearly

above the boiler as possible. It is fed from the main water supply via a ball valve, and accepts the

overflow from the open safety vent if the boiler overheats.

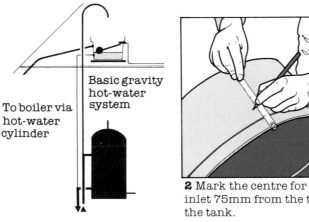

To boiler via hot-water cylinder

Basic gravity hot-water system

To boiler direct

Basic pumped hot-water system

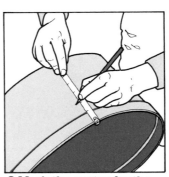

2 Mark the centre for the inlet 75mm from the top of the tank.

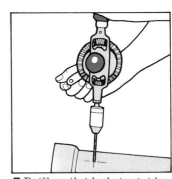

3 Drill a pilot hole to guide the hole cutter.

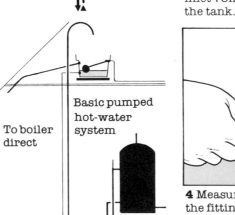

4 Measure the diameter of the fitting and select (or adjust) a cutter to fit.

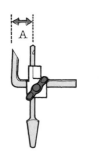

A

5 Cutters suitable for making holes in polythene tanks. Dimension 'A' equals half the hole diameter.

6 If special cutters are not available, a small drill can be used to make a larger

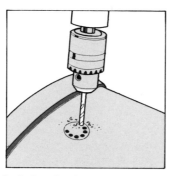

hole for the fitting. Draw around the fitting and drill a series of holes. Break out

the centre piece and clean up with a round file.

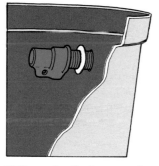

7 The polyethylene washers are placed between the backnuts and the tank itself.

8 Assemble the fitting and tighten it, using adjustable spanners. Fit other connections similarly.

9 Rig up a light in the loft but, for safety reasons, not directly over the tank.

10 Nail a platform to the joists. Remove any insulation beneath it.

11 Drill a hole through the soffit or fascia board to take the overflow pipe.

12 Rasp out the hole to size so that the overflow pipe will be an easy fit.

13 Take the tank up into the loft and connect all the pipework.

Main cold supply

Branch to feed and expansion tank

14 When the installation is complete, turn off the water, drain the pipe and connect to the main supply.

15 Add inhibitor to the expansion tank when the system is fully operational.

valves, which only operate intermittently (as is likely in a feed and expansion tank) may stick, and mains pressure is needed to overcome this.

The overflow pipe should be taken out through the soffit board, or somewhere similar, where water running out will be obvious if the ball valve is not functioning correctly.

The static water level in the tank should be just over one-third full. Gently bend the arm of the ball valve to obtain this capacity.

Finally, add the inhibitor to protect the system against rust and sludge. This job could be left for a fortnight in case any faults develop that would necessitate draining the system and losing the inhibitor.

When all the installation work is complete and the system is working, make a really careful job of insulating the tank and the pipework. A cover is required to keep the tank contents clean, but remember to cut a hole to allow the vent pipe to discharge into the tank.

An item like the feed and expansion tank in the loft tends to be forgotten once the system is working, so a job well done will pay dividends in winters to come. It is a wise move to check the tank at the beginning of each heating season, as the combination of a hot summer and a sticky ball-valve can leave the system short of water. In some cases it is also necessary to top up the inhibitor annually – follow the instructions on the pack for the inhibitor you use.

If the original cold storage tank is in poor condition (galvanized ones do corrode eventually), this is a good time to replace it. Fitting a new polythene one while you are working in the loft anyway is better than going through it all again later.

Fitting the pump

Installing a pump will often require a little more ingenuity than that required for fitting the rest of the system. Pumps are quite bulky (although they are getting smaller), and it is usually best to conceal them in some way. If you are lucky it will be installed in the airing cupboard, along with the hot-water cylinder, but more often than not it will be under the floorboards or in a specially built cupboard. If you do construct a cupboard for it, make sure that there is room for maintenance and even replacement.

The position of the pump in your installation is important, as it can have a marked effect on how the system works. If your system has been designed for you, it must be fitted where it is shown on the plan. If you have designed your own system and have followed the requirements stated in the design section and had it checked, then fit it where shown on your own plan.

In some systems the pump is fitted within the boiler casing, in which case the boiler manufacturer's instructions must be carefully followed. Certain requirements should, however, be observed. The pump should not be fitted with tees or elbows close by as this causes high resistances. Use slow bends in the pipe instead. The pump should be an easy fit into the pipework, so careful pipe fitting is required. Read the pump manufacturer's instructions carefully and follow them for a trouble-free installation.

Pumps are treated with a rust preventative at the factory, and the impeller may be a bit stiff, so rotate it a few times with a finger to overcome any resistance which could cause failure to start after installation.

One of the biggest dangers to pumps is swarf,

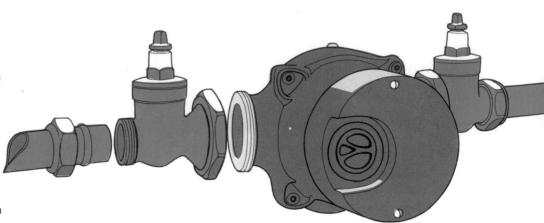

1 The pump should always be fitted between isolating valves so it can be removed for servicing (or even replacement) without having to drain the complete system. The valves are opened or shut by the use of a small spanner on the square head, or a screwdriver or allen key. Valves are available to fit either 22mm or 28mm pipe. The pump, being constructed in cast iron, is relatively heavy and must be supported securely at both ends.

2 To overcome any stickiness, rotate the pump impeller by hand a few times before installing.

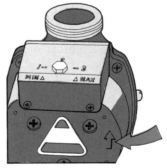

3 Check the direction of flow from the arrow cast on the pump body.

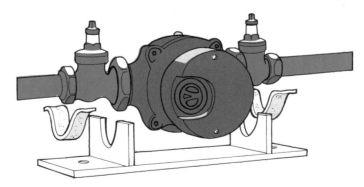

4 When installing the pump horizontally in a cupboard, make up a wooden support bracket and use insulation strips to avoid vibrations in the floor.

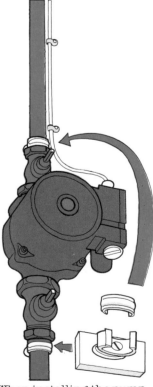

5 When installing the pump vertically, use securely fitted pipe clips for support. Fit spacers if required.

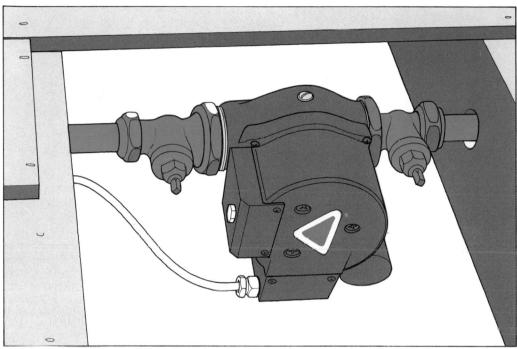

or filings from pipes, when the system is first started up. It is therefore very important to flush the system thoroughly without the pump fitted. This is done by fitting a short length of pipe between the isolating valves in place of the pump. Running water through the system by untying the ball valve and opening the drains for about fifteen minutes should clear out any debris. Make sure any control valves are in the manually open position during flushing. Shut off the isolating valves and replace the pump, reopening the valves after it is fitted.

If the pump is fitted under the floorboards, a neat, easily removable trapdoor should be fitted for access. A pump can be fitted vertically if this suits your layout, but avoid mounting it directly on to a very lightweight wall as this will amplify noises and vibrations.

When you install the pump, consider where you will be running the wiring (see the electrical section) – it is far easier to drill any small holes in the joists at this stage rather than later.

If your pump will not be used during the summer months it should be switched on every month or so for a few minutes to prevent sticking.

6 If the pump is to be fitted under the floorboards, it will be necessary to drill holes through the joists since the pump will need to be lower down to clear the floorboards. Do not cut deep slots in the joists because this can seriously weaken them.

7 A joist brace is designed for use between joists which are not wide enough for an ordinary drill to be used.

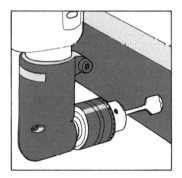

8 Alternatively, use an angle-drive attachment fitted to an electric drill.

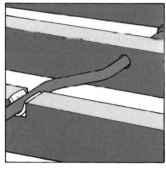

9 As soon as possible, raise the pipework to the top of the joists again.

Check:

Does the direction of the arrow on the pump agree with the water flow in the system?

Are there any elbows or tees adjacent to the pump?

Are the electrical parts higher than the water pipes?

Never run the pump dry and always make sure that the isolating valves are open.

Vent the pump of air with the pump switched off.

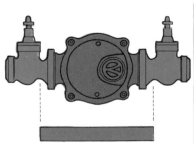

10 When the pump pipework is complete, cut and clean a piece of pipe for flushing.

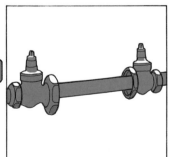

11 Remove the pump and gently ease away the pipework and slip the pump pipe into position.

12 When the system has been flushed, replace the pump, open the valves and vent any air.

Fitting control valves

No control which restricts the flow of water through the cylinder should be used with solid-fuel boilers unless special provisions are made (see pages 60-61).

These instructions cannot cover all aspects for all valves; as with all sophisticated controls, the manufacturer's instructions should be carefully read and observed.

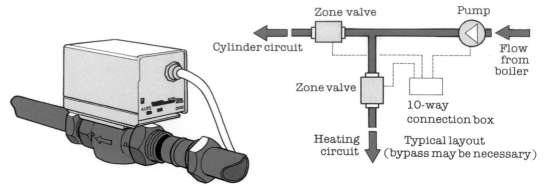

Decide where the zone valve(s) need to be fitted, according to your own plan. Remember that access is required (as with pumps). Connections are simply compression fittings to 22mm or 28mm pipe. The motor-head should not be fitted below the level of the pipe.

Zone valves and diverter valves

Being rather bulky, these valves can be fitted in a cupboard along with the pump if this suits your layout. Some types are compact enough to fit under the upstairs floorboards but the joists will need to be drilled as with pumps.

Use jointing compound sparingly on the olives of compression fittings and keep away from the moving parts of the valve. Other valves, such as safety valves, may require soldered joints such as end-feed fittings.

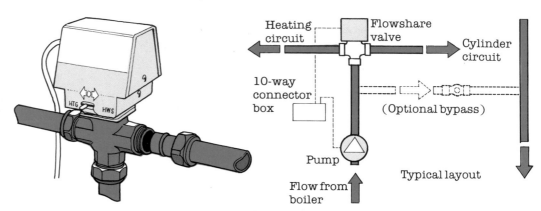

Diverter valves are fitted into the pipework using the compression fittings supplied. Access will be required for possible future maintenance.

Hot-water control valves

Hot-water control valves control the temperature of the water by shutting off the flow through the primary pipes on the cylinder circuit once the required temperature is attained. The temperature of the water at the taps is then at a lower (and safer) temperature than the heating system.

Two basic types of hot-water control valve are available: the self-contained mechanical valve and the remote sensor valve. The remote sensor is strapped to the cylinder at about one-third to half way up. If the cylinder is the pre-insulated type, the insulation should be cut away at the point of contact. On no account cut the sensor pipe; if it is too long, coil it loosely at a convenient point but out of the way.

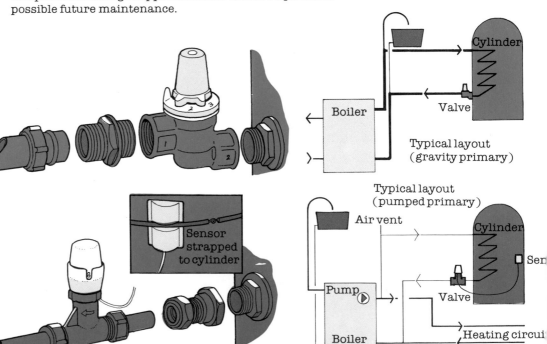

The hot-water cylinder control valve is fitted in the primary return pipe, either close by or directly to the cylinder. If a remote sensor is used, it is strapped to the cylinder about a third of the way up.

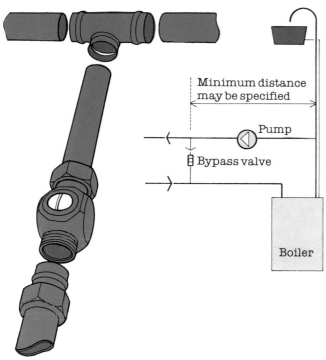

The bypass valve is fitted into a 15mm pipe connecting the flow and return pipes near the boiler. Minimum distance from the boiler may be specified by the manufacturer.

Adjustment of the bypass valve

Boilers that require a bypass (usually low-water-content, wall-hung boilers) will have specific instructions included, but the basic procedure is as follows:

1 Adjust the pump setting to satisfy the system's needs, but ensure the flow is sufficient to meet the minimum requirements specified by the boiler manufacturer.
2 Turn the boiler thermostat up to its maximum setting.
3 Fire up the boiler with the bypass valve fully closed and with all the radiator and control valves fully open (full load).
4 Adjust the system to minimum load. This will be either the cylinder circuit only, or the heating circuit with only one radiator operating.

5 Gradually open the bypass valve until the boiler operates quietly and at all flow temperatures.

Once the valve is adjusted it should be left at this setting. If a gate valve is used, a tie-on label with a warning should be attached to it.

Safety valves and drain valves

Obviously only those controls designed into your own system are fitted, the exceptions being the safety valve, which should always be incorporated, and the drain valves which are necessary for emptying the system (or part of it).

One drain valve should be fitted at the lowest part of the pipework and this will normally be sufficient if the pipes from the boiler fall to here. If the boiler is on an enclosed loop – for example a wall-hung boiler with pipes rising to the first floor before diverting to their various circuits – then a drain will be fitted to the lowest boiler tapping. Looking at your isometric drawing will usually suggest the logical drain points. If you use a drop-loop system for radiators, a drain will be required at each one.

Air valves

On radiator circuits the normal air valves will vent any trapped air, but on a pumped cylinder circuit an air valve will be required at the highest point of the pipework. On a gravity cylinder circuit the open vent will allow air to escape.

Balancing valves

A balancing valve may be needed to restrict the flow on a pumped cylinder circuit. All that is required is a screw-regulated mini-valve (as shown for the bypass), or a gate valve adjusted to the required amount, fitted on the cylinder-return to the boiler.

Check valve
This type prevents unwanted gravity circulation when the pump is off.

Safety valve
The safety valve should be fitted on the flow pipe as close to the boiler as possible.

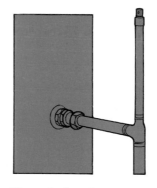

Manual air valve
Fit on the highest part of a pumped cylinder circuit. Extend the pipe about 200mm above the tee.

No valves, with the exception of safety and drain valves, should be fitted on the open vent and cold feed pipes. Allow for continuous water circulation with low-water content boilers via a bypass where shut-off valves are incorporated.

Make sure the pump will not be operating against shut-off valves.

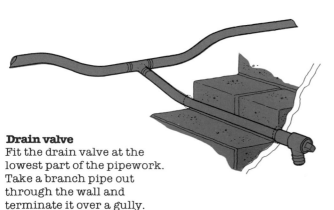

Drain valve
Fit the drain valve at the lowest part of the pipework. Take a branch pipe out through the wall and terminate it over a gully.

The wiring

Using a mains tester: the screwdriver tip is placed on a terminal and the top of the tool touched with the finger (there is no danger of shock as a resistor is built into the handle). A 'live' circuit will light up the neon lamp in the transparent handle.

Tools and materials

1 Torch: one that stands firmly on its own base is the most useful type.

2 Trimming knife: useful for stripping away outer layers of insulation on cables.

3 Mains tester: a safe tool for determining if mains current is present at a terminal.

4 Wire strippers: two types of wire stripper which can be adjusted to the thickness of the copper core, avoiding the danger of cutting right through the cable.

5 PVC insulation tape: useful for tidying cable runs by taping groups of cables together. Do not use insulating tape for joining cables: use proper junction boxes rated for the load.

6 Cable clips: for securing cables neatly to joists and walls. The pins are in fact small masonry nails. Cable should never be left exposed on room walls; use the special PVC trunking for safe protection.

7 Screwdrivers: two sizes will be needed — a large one for undoing stiff screws on electrical items, and a small one, slim enough to penetrate the plastic connections on some types of junction boxes.

8 Cable: the cable used for central-heating installation work should be PVC 1.5mm² single strand, twin core and earth.

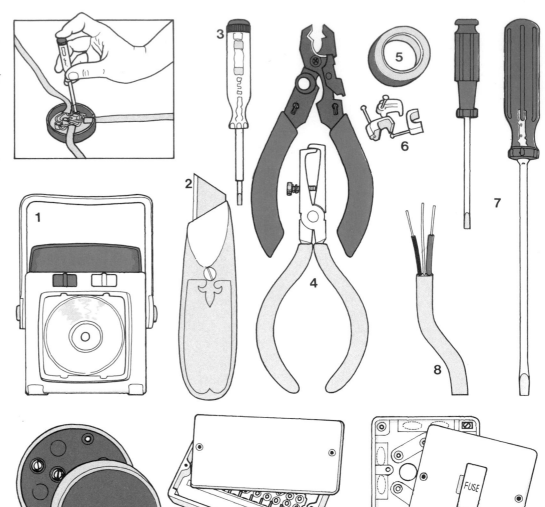

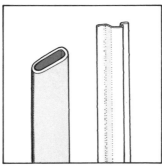

Junction boxes are used when breaking into a ring main to make a mains connection for an installation. Use a 30-amp junction box.

Ten-way connection box: used as a connecting point between the various electrical items in a system: pump, boiler and thermostats, etc.

Double-pole fused isolator: this can be switched or unswitched. Removing the fuse renders the system 'dead'.

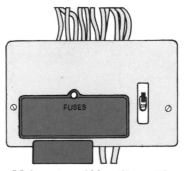

PVC conduit and PVC or galvanized channelling should be used where cables are buried in plaster.

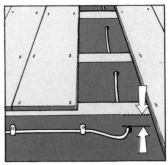

Make yourself familiar with your own consumer unit near your electricity meter.

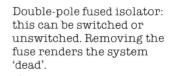

When running cables under floors, drill holes at least 50mm below the top of the joist.

Running cables in walls

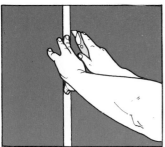

1 Cutting through a spare piece of wallpaper for later matching will cause minimum disturbance to decorations.

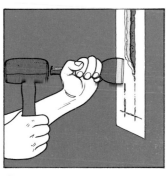

2 Chop out sufficient depth for the cable conduit or channelling. Cut into the brickwork if necessary.

Switch off the electricity

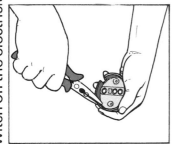

1 Break out three entries for the cables: two for the ring main, one for the spur.

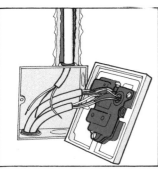

4 Alternatively, connect and run a cable from the back of a convenient socket.

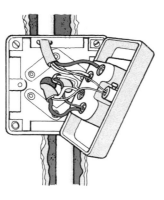

3 To route a cable through a ceiling, push a thin screwdriver up through and locate the end.

A convenient arrangement for a programmer, a room thermostat and a fused isolator. The wiring would run through to the first floor for the pump, the boiler, the cylinder thermostat and the zone valves.

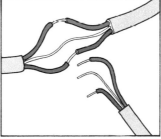

2 Strip off the outer PVC covering from the ring main cable and separate the three cores.

5 When the system has been tested fill the channel and allow to dry thoroughly.

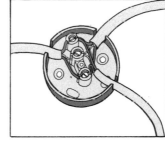

3 Cut the inner insulation away at three points and connect the spur cable with the ring main.

6 Coat the filled area with size before pasting the previously cut wallpaper into position.

Electrical safety

Mains electricity can kill — treat it with the utmost caution. Do your wiring jobs during daylight hours and switch off the entire house electricity at the consumer unit before touching any mains wiring. Double-check by plugging an electric drill into a socket and switching lights on and off. It is best not to rely on isolating part of the circuit by removing a fuse, as a mistake could have disastrous results. If electricity is needed, say for a drill, plan your work so you can do all the drilling first.

Always do electrical work in reverse, working back towards the mains connection. This way you can do all the wiring in complete safety, checking your connections, and when satisfied, switch off the electricity and make the final connection. The worst you can do then is to blow a fuse. If a fuse does blow, the cause should be found and rectified. If you take a cool look at the problem the fault will usually become obvious. It may help to make a small sketch plan of the circuit to see where you may have gone wrong.

The final connection to the fused isolator can be conveniently made via the ring main or from the back of an existing socket. If your new wiring runs close to the consumer unit you may find a spare outlet that you can use. A rubber grommet should be used where the cable passes through the socket box. If using the ring main, don't cut it, just bare the cables and push them into the terminals, securing the spur cable ends on top.

All responsible people in the house should know how to switch off the electricity supply.

Warning: In some countries, including Australia and the USA, wiring must be done by a qualified electrician.

Microbore installation

Roughly half the work in a microbore installation will differ from the standard small bore system. Some parts of the installation do not differ:

1 The boiler. No difference, although low-water-content boilers go particularly well with microbore installations.
2 The hot-water cylinder. No difference, unless a conversion unit is being used.
3 Controls and control valves. No difference.
4 The pump. No difference in installation, but the pump head will usually need to be higher and the pump and pipe sizing carried out more carefully. (See page 56.)
5 The feed and expansion tank. No difference.

The main difference is between the radiators and the distribution units known as manifolds. One manifold is usually required for each floor: a bungalow with a large floor area may be divided between two manifolds. Manifolds have from eight to eighteen connections (both flow and return are on the same unit) and have either 22mm or 28mm mains connections. The mains pipework is run directly from the boiler to the manifold via the pump and any control valves.

Decide which side is to be the flow and which the return, and mark the manifold accordingly.

All outlets are 10mm and, according to the radiator load and/or the length of pipe, either 10mm, 8mm or 6mm pipe is used. With the two smaller pipe sizes, a reducing set will be required at the manifold and the double-entry radiator valve.

Microbore pipes can be run through holes drilled about halfway down the joists or clipped neatly with special pipe clips where they run alongside them.

Any unused outlets on a

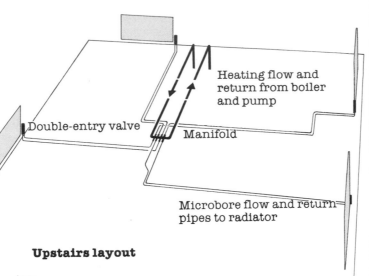

Heating flow and return from boiler and pump

Double-entry valve

Manifold

Microbore flow and return pipes to radiator

Upstairs layout

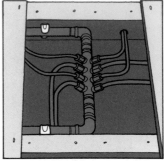

A typical underfloor manifold installation: the mains pipes can be clipped to adjacent joists.

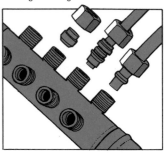

Reducing sets are used for smaller pipe sizes. Unused outlets are blanked off.

The usual layout in microbore systems is to have a manifold to serve each floor. It is important to situate the manifold as centrally as possible to equalize the pipe loops to each radiator because this helps to balance the system. The manifolds are served by either 22mm or 28mm pipes, according to the heating loads imposed on that part of the system. All outlets are 10mm.

1 Before feeding pipes under floors, squeeze the ends to keep out dirt.

2 The ends can be drilled and a cord attached for pulling through confined spaces.

3 Straighten out pipe as much as possible before installing it.

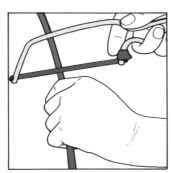

4 Cut microbore pipe with a hacksaw, not with a pipe cutter.

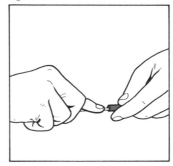

5 Smooth off any burrs from the inside end of the pipe with a round needle file.

6 Use the special bending spring (which fits over the pipe) for making small radius bends.

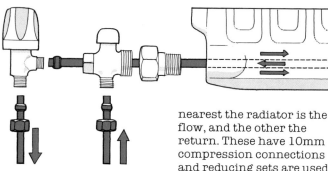

The valve shown here has a balancing facility, but installation is similar for all double-entry radiator valves. The connection

On double radiators where it is impossible to insert a straight pipe, use a flexible one instead.

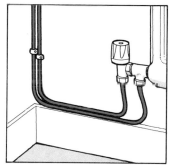

nearest the radiator is the flow, and the other the return. These have 10mm compression connections and reducing sets are used for smaller pipe sizes. A length of 10mm pipe with an olive is fitted into the radiator to ensure good circulation.

On drop systems, pipes can be clipped neatly and unobtrusively to the wall.

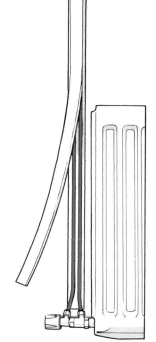

Alternatively, pipes can be hidden in the type of plastic channel used by electricians.

manifold are blanked off using special, self-sealing blanks. It is a good idea (especially on ground floors) to have a spare pair of blanked-off outlets because, if you extend your house at a later date, a new radiator can be added easily.

Radiator valves

The valves used in a microbore installation have both the flow and return at the same end of the radiator, so the other radiator tapping must be blanked off or a drain valve fitted if a drop-loop system is used.

Although the microbore system is claimed to be self-balancing, because separate pipe loops are used to serve every radiator individually, a double-entry balancing valve is also available.

Valves can be rotated to any position, which is particularly useful where radiators are served by dropping the pipes from above.

Single-panel radiators require a 200mm length of 10mm pipe inserted to provide return circulation. Double-panel radiators require a flexible insert to enter either panel. A longer pipe is necessary (up to two-thirds the length of the radiator) on low radiators (up to 400mm high) and those over two metres in length.

If conventional or thermostatic radiator valves are used instead of double-entry valves, a reducing coupling, 15mm x 10mm, is used to connect the pipe to the valve.

Take care when handling microbore pipe. It is soft-temper copper and can be easily flattened, for example by dropping a coil of it on a hard surface such as a concrete floor. Don't overtighten unions: microbore pipe seals easily and does not require any force.

Installing a conversion unit

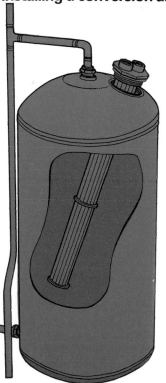

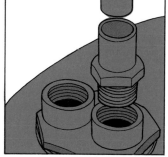

1 Use a ¾in BSP male to 22mm soldered or compression fitting for connecting primary pipework.

2 Fit an air vent to either connection where the hot-water primary is pumped.

The unit shown here is a 'Micraversion'. This has a series of tubes brazed together to give a good distribution of heat into the secondary water. With this method, a direct cylinder (in good condition) can be converted into an indirect cylinder fairly cheaply.

The unit is screwed into the immersion-heater boss and the pipework run directly from the boiler. This is suitable for gravity (with the provision for an open vent) or pumped systems. It does not necessarily have to be combined with a microbore heating installation. (See pages 55 and 57 for layouts of both pumped and gravity systems.)

Order of work and system checklist

1 Send off for catalogues of boilers and materials from the large heating-supplies warehouses, or visit a local stockist and discuss your requirements with them. Even if you are buying the whole system by mail order, it is worth locating your local supplier as you will almost certainly underestimate the quantity of 15mm pipe (which comes in 3m lengths), and 15mm elbows required. It is also useful to be able to discuss any problems with them, and they will usually exchange a fitting bought in error, provided it has not been used.

2 Measure all the rooms and windows for heat loss, and make notes on the house structure. These measurements will still be needed even if you are not designing the system yourself, since the designer will need them for his or her own calculations.

3 Design your installation. Read the design section carefully until you understand fully the method of central-heating design and can apply it to your own house.

4 When your plans are complete, carefully list every item necessary to install the system. Include tools if these are needed. (Use the itemized checklist on pages 106-7.)

5 Decide on the positon of the boiler and fit it exactly according to the manufacturer's instructions. With solid-fuel and gas room-heaters, the boiler could be installed towards the end of the work schedule, as boiler pipework needs to be buried in the chimney breast and restoration cannot be completed until the system has been tested for leaks. Fit the flue-liner, if required, at this stage.

6 Hang the radiators in their respective positions and loosely fit their valves.

7 Fit the hot-water cylinder. If the original cylinder, or its position, is to be used, then this job could be left until later.

8 Refer to your plan and run the pipes.

9 Fit the necessary control valves and fit the pump temporarily. A short piece of pipe will be needed in place of the pump for flushing the system clean.

10 Fit the feed and expansion tank in the loft and connect to the boiler pipework. Run the cold feed to the feed and expansion tank, and adjacent to the house rising main, but do not connect it at this stage.

After the whole system has been checked and any leaks cured, open the drain valves and allow the system to run through for about ten minutes to flush away any metal filings.

Close the isolating valves, remove the pump pipe, and refit the pump permanently. Open the isolating valves and bleed the pump by unscrewing the bleed screw on top of it.

Wiring

Complete all the electrical wiring according to the wiring diagrams, but leave the fuse out of the fused isolator so that the system is not 'live'.

Commissioning the boiler

For gas and oil installations call in an approved fitter; for solid fuel, light a small fire according to the manufacturer's instructions. Depending on the arrangements made for boiler commissioning, the fitter may check the system, balance the radiators and

11 Check thoroughly all the pipework and fittings, making sure you have not forgotten to solder any.

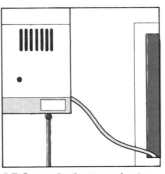

13 Open the bottom drain cock and the boiler drain cock (if fitted) and run hoses to outside drains.

15 Manually open any control valves, if these are fitted.

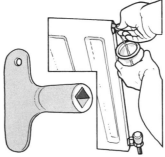

17 Close the drain cocks and bleed the radiators; start with the lowest and work up to the highest.

Main cold supply

Branch to feed and expansion tank

12 Turn off the main cold supply and connect up the feed and expansion pipe.

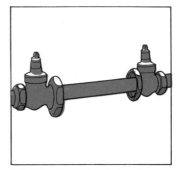

14 Fit a length of pipe in place of the pump and make sure the isolating valves are open.

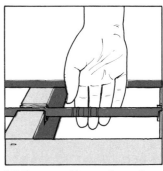

16 Turn on the main water supply and quickly check for leaks at all joints and connections.

When checking for leaks, a slight dampness is to be expected due to condensation as the pipes fill with cold water. With leaking compression joints, simply tighten the joint with spanners; with soldered fittings there is no alternative but to drain the whole system and dry the joint with a blowlamp before applying more solder.

adjust the pump and any controls, or the installer may have to complete these jobs. If so, the following tips will be useful:

Air locks in radiators

First, try bleeding the radiator. If this does not clear it, turn off all the other radiators and switch on the pump. Turning the pump up to its maximum setting for a short period will usually clear it.

Adjusting controls

Make sure all zone valves are in their electrically operating position and not in the manually open position. Turn room thermostats and cylinder thermostats to their intended setting and turn the boiler thermostat to its maximum setting.

Balancing the system and pump adjustment

Two pipe thermometers are required (it may be possible to hire these). One is clipped to the boiler flow pipe and the other to the return. When the boiler has been operating for an hour or so, the temperature difference between the two pipes should be 11°C. If it is greater than this, increase the pump setting to a higher number. This will decrease the difference. Allow the system to settle and transfer the thermometers to the flow and return pipe of each radiator in turn. Adjust the lockshield valve of each radiator to obtain the 11°C difference.

If any radiator is noticeably cooler and all control valves are functioning correctly, increase the pump setting to overcome this resistance. Generally, the pump should be adjusted to the lowest setting that will operate the system satisfactorily.

Minimum flow requirements

Some low-water-content boilers state a minimum flow rate through the boiler. After you have adjusted the pump as described, check the regulator setting you have just made against the pump performance graph (supplied with the pump) to see that it meets the minimum flow required.

If thermometers are not available, adjust the radiators by touch. You will usually find the radiator furthest from the boiler is the coolest (the lockshield valve should be fully open here), and the radiator nearest the boiler is the hottest. Progressively shut down the lockshield valve here to increase resistance.

It may be difficult to balance the system perfectly during the warm summer months, so you will have to wait until winter.

Warning

Do not start filling the system late in the day because problems may arise. Set a whole day aside and start in the morning. In cold weather do not fill the system until you are sure that the commissioning fitter (for gas or oil) is able to come, as the pipework is liable to frost damage because it is not yet lagged.

Finishing off

This is probably the most satisfying part of the whole job. The hard work is behind you, the system has been checked and is working properly and you have added extra value to your home, as well as ensuring comfortable winters for the years ahead.

Make sure, before you replace the floorboards permanently, that all the pipes under the ground floor are well protected with no gaps in the insulation.

Floorboards are best screwed down rather than renailed, and this will also help to cure creaking if this is a problem with old boards.

As well as insulating the feed and expansion tank, fit it with an insulated removable lid. Make a hole to allow the open vent to discharge into the tank. When the system is functioning satisfactorily, add the inhibitor to the feed and expansion tank.

1 Carefully insulate every part of exposed pipework in unheated areas.

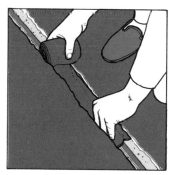

2 Protect pipes in concrete floors with Densotape before making good the floor screed.

3 Fit a really thick, good-quality cylinder jacket to minimize heat losses from the cylinder.

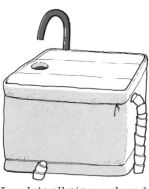

4 Insulate all pipework and the feed and expansion tank in the loft.

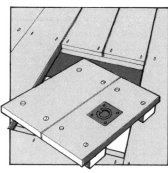

5 Make a simple access trapdoor for the pump and control valves.

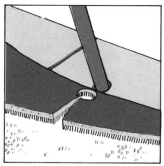

6 Refit carpets and floor coverings around the new radiator pipes.

Radiators for difficult areas

Although the best site for a radiator is under a window in a room, there may be rooms (particularly in older houses) where fitting problems occur. One of the first things to be considered is whether the radiator will be obstructed by the furniture — this is one of the reasons why the window is a good place because large items of furniture are not usually placed there.

Curved bay windows
Curved bay windows present an additional expense when buying radiators, but it is worth spending the extra money on a curved radiator because, besides being a window and subject to an extra heat loss, a bay is exposed all round and therefore a particular cold spot in winter. Large central-heating suppliers will curve a radiator to your measurements. The radiator should be sized as described in the design section and a template made as a guide. Take an old roll of stiff wallpaper, cut a length to fit the bay and, by trial and error, trim a curve to fit the bay wall exactly. Take time to get it right — you won't get a second chance. Match the wall, not the skirting board, since this will be a different radius.

Square bay windows
Square or hexagonal bays can be fitted with two or three narrow radiators piped as shown. From the range of shapes and sizes of single, double and convector-panel radiators available you should be able to make up the output required for that room.

Convector radiators
Much more expensive than panel radiators, convector radiators (as opposed to panel-convector radiators) have finned pipes contained in a decorative, sometimes hardwood, casing and are fitted with a fan controlled by its own thermostat. They produce a high output from small dimensions and can be used where a panel radiator would not fit.

They are particularly useful for kitchens — the slim vertical types can be fitted flush with the front of the kitchen units: down-draught convectors can be fitted over doors; and the ingenious kickspace convector uses space that would otherwise certainly be wasted.

Full fitting instructions are supplied with all convector radiators.

Using pipe heat
In a bedroom with a wall prone to dampness, for example behind a wardrobe where air cannot circulate, use can be made of pipe emission by routing the pipes to the radiator along that wall under the floorboards, where a gentle heat will rise to counteract the problem. Any dampness due to structural faults should be tackled at source.

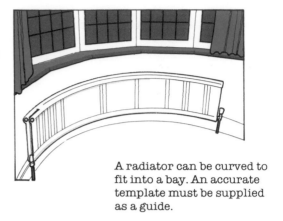

A radiator can be curved to fit into a bay. An accurate template must be supplied as a guide.

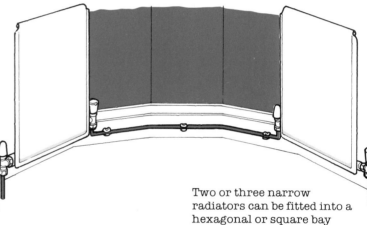

Two or three narrow radiators can be fitted into a hexagonal or square bay and linked.

Tall narrow fan convectors provide a high heat output where space is limited.

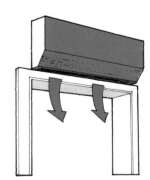

Down-draught fan convectors are useful for kitchens where wall space is not available.

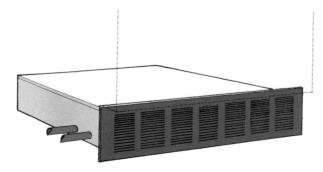

A kickspace convector is fitted into the base of a kitchen unit and controlled by a remote switch. It provides heat where it is most beneficial — at floor level.

Fitting a new boiler

Installing a new boiler in place of one that has worn out can range from a simple changeover job to an almost complete redesign and refit of the whole existing central-heating system.

Selecting the boiler output

The first requirement is to ascertain the output required from the new boiler. If the system is to remain basically the same and the previous boiler was properly sized and coped sufficiently well before, you can take the rating of the old boiler. If you don't know this, write to the manufacturer. If the old rating is in Btu's it will need to be converted to kilowatts, divide Btu's by 3412 to give kilowatts. Once you have

decided on a boiler, try to get a copy of the installation instructions beforehand and study them very carefully so that any likely problems can be anticipated.

If you are adding radiators, their outputs must be included in the total boiler output requirement. If the changeover is fairly straightforward, and the pipe connections from the old boiler and the new boiler not too different, the work will be minimal. Identify the various pipes, list carefully all the materials and tools required and plan the work to cause as little disruption as possible.

If it is a gas boiler, you are not allowed to disconnect or connect to the gas supply. Get an approved installer to do this part of the work.

If the work entails

anything more than what has been outlined above (for example, changing the boiler site from one room to another), plans should be drawn up as shown in the design section.

Designing for a new boiler

First, assuming the radiators are in good order and in the right places, draw these on the plan. Next draw the boiler in its new position, and the hot-water cylinder. Mark in the pipe runs, the open vent and the feed and expansion tank.

Preparatory work

Once you have drawn the plans to your satisfaction, start doing some exploratory work. Go up into the loft and find the existing feed and expansion

tank and make rough drawings with measurements. Without disrupting the household too much at this stage, try to find where the pipes run under the floors.

If possible, take measurements and note the pipe sizes. Take particular note of exactly where the pipe sizes increase in diameter as this is important for pipe sizing calculations. Now make a rough drawing of your old system and compare it with the new layout. It is likely that, with slight alterations to your new drawing, most of the existing pipework can be left undisturbed.

The old pipework may be in imperial sizes, but this is no problem as adaptors are readily available to connect metric to imperial pipes.

Decorating behind radiators

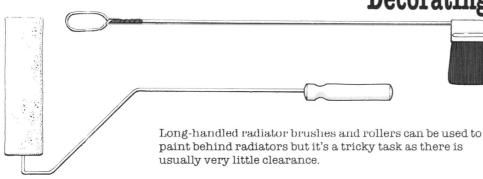

Long-handled radiator brushes and rollers can be used to paint behind radiators but it's a tricky task as there is usually very little clearance.

There are two ways of tackling this problem — one is to use a long-handled brush or roller and the other is to remove the radiator itself, which is not as difficult as it may seem.

Removing a radiator

1 Close the handwheel valve, then the lockshield valve, counting the turns as you do so.
2 Place containers at either end of the radiator to catch as much water as possible, and undo the large nuts connecting the valves to the radiator.
3 Gently lever the valves away from the radiator and lift it off its brackets. After decorating replace the radiator, following the instructions for hanging radiators. Open the handwheel valve fully and the lockshield valve by the same number of turns as it was closed, then bleed the radiator.

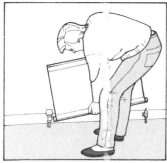

Remove the radiator by undoing the nuts at either end. Be prepared for spillage.

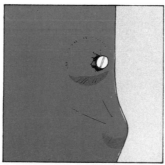

An easy way to find bracket positions is to paper over the screws.

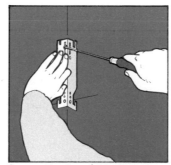

Screw the brackets tightly to the wall when the decorating is complete.

Faults in central-heating systems

Various faults can develop in a heating system, some serious and others which you may be able to correct yourself without having to consider calling in an engineer.

Leaks
A small leak at a joint can be cured on a compression fitting by gently tightening the nuts. On a solder fitting it will be necessary to drain the system and heat the joint with a blowlamp and add more solder. A threaded joint at a boiler, cylinder or radiator can be tightened down but if this fails, drain the system and remake the joint.

Cold radiators
A radiator which is cold at the top and hot at the bottom will probably just need bleeding.
A radiator which is cold at the bottom but hot at the top

probably contains sludge caused by corrosion. Remove the radiator and thoroughly flush it through with a hose.
More than one cold radiator may mean sludge throughout the system or the pump not working properly. Locate the pump; if it's noisy it will probably need replacing.
If radiators constantly need bleeding, check the feed and expansion tank in the loft to see that the water level is correct and that the ball valve is not sticking. If it is, you will have to replace or re-washer it.
Another cause of cold radiators throughout the house could be that the room thermostat may gave accidentally been turned down.

Boiler faults
These are the most serious of all the faults you may

encounter, and a serviceman will probably need to be called; but check first for air-locks, lack of water in the feed and expansion tank, and blocked flues.

Electrical failure
Check the fuse in the

isolating switch and the fuse in the boiler (see the boiler manufacturer's literature). If a fuse has blown, the cause must be found and rectified. Check for broken or loose wiring, *switching off the electricity first*: trace the wiring back to the junction box.

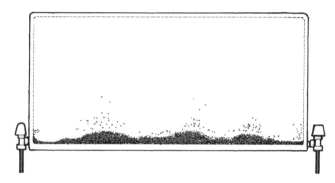

If an inhibitor was not added to the system, radiators can corrode internally, and a build-up of sludge causes them to become cold at the bottom and eventually fail.

Extensions

If you are having an extension added to your home and you already have central heating, you should plan at an early stage the positions of the new radiators and the pipes to serve them. One or two additional radiators shouldn't affect your boiler too much, but more may require a larger one.
The proposed room (or rooms) should have their heat losses calculated and the radiators selected as described in the design section. Once the walls are up and the window frames fitted you can position the new radiators and mark where the pipe connections will come up through the floor. As the floor screed is one of the last jobs to be completed in an extension, ask the builder to lay battens across the floor to where the

new pipework will connect to the existing pipes. The pipework serving the new radiators can be planned to connect either to the existing 15mm pipe, where a room is extended, or into the 22mm or 28mm pipework serving other radiators where a separate room (or more) is planned.
It's not advisable to hang radiators before building work is complete, since they will get splashed with mortar and plaster.
When the system is refilled, all the radiators will need to be bled; and balancing will be necessary as the dynamics of the system will have altered.

Note: check with the local Water Board to see if they object to central-heating pipes being buried in floor screeds.

Replacing a radiator

Replacing a radiator because of corrosion or other problems is not difficult, particularly if the new radiator is the same length. Any carpets should be removed and the floor covered with polythene and newspapers to avoid staining.

1 Fully close the handwheel valve, then the lockshield valve, counting the number of turns as you do so.

2 Undo the large nuts close to the radiator and spring the pipes slightly to release it. Use containers to catch the water and lift the radiator off its brackets.

3 Remove the tails from the old radiator and fit them to the new one, using PTFE tape on the threads. If the valves or tails are not usable, new lockshield and

handwheel valves will be required. In this case, the whole system must be drained to allow for the changeover.

4 Fit the new radiator, and new brackets if the old ones don't fit. See also the instructions for fitting radiators given on pages 68-9.

5 Open the handwheel valve fully and the lockshield valve by the number of turns previously counted, and bleed the radiator.

6 If the new radiator is longer or shorter than the old one, extensions to the pipework under the floorboards will be necessary. Study the instructions shown in Removing floorboards on pages 78-9 and Running the pipes on pages 80-81.

Replacing a pump

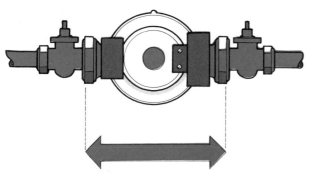

Measure the pump from flange to flange so that a replacement can be purchased which will fit without too much disruption to the system.

Close the isolating valves right down and place a shallow container beneath to catch the water.

Undo the large flange nuts to release the pump.

Where pipework needs to be altered, saw through the pipes at a convenient spot.

Old pipework may be in imperial sizes, so use an adaptor to connect.

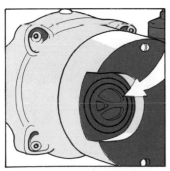

Adjust the regulator on the new pump to the required setting.

Finally, bleed the pump of any trapped air. Do this with the pump turned off.

Old pumps probably cause more problems in heating systems than almost any other single item, and they will usually need replacing well before the rest of the system is worn out. Pumps are neater and quieter than ever before, and use more advanced materials for their moving parts.

Pump replacement is a simple, straightforward job if the new pump is similar to the old one. More work is involved when the pump is very old and perhaps fitted without isolating valves, because alterations to the adjacent pipework will be necessary.

The replacement procedure is as follows:

1 Measure the distance between the pump flanges as shown on the left.

2 If a regulator is fitted, note the setting. Note also the make, model and number of the pump.

3 Take these details to a heating supplier and try to obtain the same or similar model. The supplier will have pump performance tables from which he will be able to supply a pump of the same performance even if it is a different make.

4 Switch off the electricity at the consumer unit. Trace the pump wiring back to the connector box and make a simple sketch of which wires go to which terminals. Disconnect the wiring.

5 Close the isolating valves on either side of the pump. Where no valves are fitted it will be necessary to drain down the system.

6 Hold the pump body and undo the large flange nuts.

7 Ease the pipework away from the pump and remove it. There will be some water in the pump, so be prepared for this.

8 Clean the pipe connections of any old jointing compound, PTFE tape or hemp. Renew the fibre washers if these are fitted.

Note the direction of flow from the arrow stamped on the pump body and, where the new pump is a similar fitting to the old one, follow the instructions on pump installation.

Where pipework needs to be altered, or isolating valves fitted where there were none before, take careful measurements and saw off the pipes at the required distance after draining down the heating system. Fit the isolating valves and pump. If pipework requires altering, find a convenient place where there is room to work, saw through the pipes and connect the new pipework.

On an old system the pipework may be in an imperial size, in which case you will require adaptors to connect the old pipe to the new.

Where a new pump is of a different length to the old one, spacing connectors are available to fit between. On no account fit a new pump without isolating valves.

9 Make sure the electricity is turned off and make the electrical connections according to the sketch plan made earlier.

10 Having fitted the pump, open the isolating valves fully. If the system was drained down, refill and bleed the radiators as the system fills.

11 Bleed any air from the pump and adjust the regulator to the required setting.

12 Switch on the electricity supply and any other electrical controls to operate the system.

13 If the system was properly balanced before, it should not require any further adjustment.

Replacing a cold-water storage tank

If the cold storage tank in your loft is an old galvanized one, it may be showing signs of corrosion; replacement now while installing central heating will be much more convenient than waiting until it leaks. The fittings are all the same as those described for the feed and expansion tank.

The new polythene tanks are cheap and easy to handle and some can even be folded without damage to pass through the loft trap door.

Depending on the system used in your house, one or two distribution pipes may be fitted. In older properties all the cold-water taps and the WC cistern come directly off the main. In this case there will be only one distribution pipe coming off the cold storage tank and this will serve the hot-water cylinder alone.

In modern houses only the kitchen tap will come directly off the main – all other taps and WC cisterns will be fed from the cold storage tank. In this case there will be two distribution pipes, one for the hot-water cylinder and another for the cold taps.

Unlike galvanized tanks, polythene ones require solid support underneath or they will distort. 19mm plywood or blockboard nailed to the joists ensures a good base.

Removal of the old galvanized tank may present a problem. It may have been installed before the roof was completed and is unlikely to go through the trap door. It can be cut into pieces with a hacksaw or an electric jigsaw, but it's a tedious job so (if there's room) move it to a corner of the loft and forget about it.

The pipework and the tank should all be insulated.

1 Prepare the new tank; measure and cut the holes and fit the pipe connectors.

2 Turn off the main water supply or the isolating valve (if fitted).

3 Disconnect the old tank after draining the hot-water system.

4 Fit a new base to support the new tank.

5 Position the new tank and make all the connections.

6 Check for correct functioning by opening the isolating valve or by turning on the main supply.

7 Gently bend the ball-valve arm until the water level stops at about 25mm below the overflow outlet.

8 Fit the insulation but do not insulate under the tank.

Schematic layout for a cold-water storage tank

Open vent from HWC 22mm

Overflow 22mm

Cold-water storage tank

15mm

Rising main

Cold feed to HWC 22mm

22mm

All pipe sizes are minimum

Isolating valve

Hot-water cylinder

Primary heating pipes

Upstairs cold taps

Upstairs hot taps

Drain

Downstairs hot taps

To cold taps and cisterns

A circular polythene tank can be folded to pass through a small trapdoor.

A jigsaw with a metal-cutting blade can be used to cut up a galvanized tank.

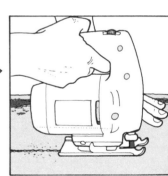

Tank arrangement in the loft

A Open vent from HWC
B Open vent from boiler
C Cold storage tank
D Feed and expansion tank

E Rising main
F Overflow pipes
G Cold feed to HWC
H Cold feed to boiler

Table of tank sizes

Total capacity/ Working capacity (litres)	Size (mm)
182/114	710 × 510 circular
227/155	610 × 610 × 610
273/182	760 × 610 × 610

Insulation

10% draughts through doors, windows and chimneys

20% through roof

10% rough dows

25% through walls

10% through floors

In the average uninsulated house, up to 75 per cent of the heat produced is lost through the house structure.

Good insulation can dramatically reduce the rate at which heat is wasted, and will therefore cut heating bills by a significant amount.

Walls
The greatest loss is through the walls, but if those in your house are solid the remedies are not cheap or easy. Exterior insulation is possible but it is very expensive, and interior insulation requires a lot of work.

Cavity walls are easier to deal with. Three types of cavity infill are available: urea formaldehyde foam, polystyrene beads and blown mineral wool or fibreglass. They must all be installed by specialist contractors.

The roof
Insulating the roof is easier and can be done by a home owner. There are two basic types of insulation, both very efficient: fibreglass blanket, in rolls 100mm thick; and loose vermiculite granules spread between the joists to a depth of about 150mm.

Draughts
This is where the most cost-effective savings can be made, and a weekend spent draught-proofing will soon repay the effort.

Since conventional-flued boilers, room-heaters and open fires all draw their air for combustion from the room, make sure that permanent ventilation is available.

Floors
Fitted carpets with a thick underlay are a good form of insulation. Floorboards with rugs will require careful draught-proofing.

Windows
Secondary double glazing will reduce heat lost through glass, but the draughts should be tackled first.

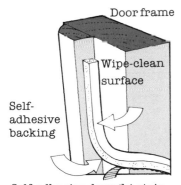

Door frame

Wipe-clean surface

Self-adhesive backing

Self-adhesive draught strip comes in rolls, is effective, inexpensive and easy to apply.

Door frame

Sprung strips in bronze or nylon fitted to door frames seal relatively large gaps.

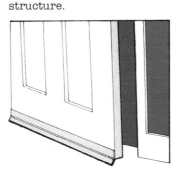

Various types of draught-excluder are made for fitting to the bottoms of doors.

Fibreglass in rolls is the most popular form of loft insulation.

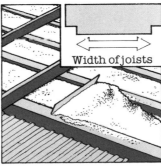

Width of joists

Polystyrene or vermiculite granules are poured straight from the bag and levelled off.

Thin polystyrene wall veneer can be applied before wallpapering to insulate cold walls.

Chimneys and flues

Just because a chimney functioned well with the fire or boiler that was used previously, it does not mean that it will be exactly right for a new high-efficiency boiler, and the flue interior must be of the correct shape for the appliance. If the house is in a built-up area with higher buildings nearby, the chimney may never actually work properly and a balanced-flue boiler may need to be installed instead.

Problems concerning the condition of the inside of old brick chimneys can be overcome by having them lined; this is expensive and not a DIY job, however, unless flexible liners are to be used.

There are some ingenious systems on the market — in one type an inflatable rubber former is inserted into the chimney and lightweight concrete is pumped into the surrounding air space. When the concrete has set, the former is deflated and withdrawn; the result is a smooth round flue of the correct diameter for the appliance.

Another type uses circular block liners lowered section by section through the top of the chimney; where there are bends, the chimney breast must be broken into. The surrounding area is filled with lightweight concrete to complete the job.

Other flue systems are available for internal use, made in twin-walled stainless steel and featuring various ingenious fittings for passing through a ceiling and out through the roof.

You should discuss special problems with the company supplying your heating materials, as they carry stocks of leaflets describing these systems in detail. These leaflets will also show the correct type of chimney to select for the fuel of your choice.

Position of balanced flues for gas boilers (minimum distance mm)

Terminal position	Natural draught	Forced draught
Directly below an openable window or other opening e.g. air brick	300	300
Below a gutter	300	75
Below house eaves	300	200
Below a balcony	300	200
From a drain pipe or a soil pipe	600	75
From internal or external corners	600	300
Above the ground or a balcony	300	300
From a surface facing a terminal	600	600
From a terminal facing a terminal	600	1200

If a balanced flue terminal is fitted closer than 600mm of a plastic gutter, the gutter should be protected by an aluminium shield 1.5m long.

A guard should be fitted if the terminal is less than 2m above the ground or a balcony.

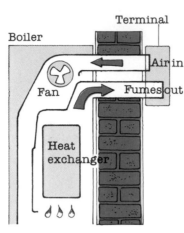

A forced-draught balanced flue works by drawing air for combustion from outside through one duct and exhausting the waste fumes through another.

There are minimum recommended distances for the siting of balanced-flue outlets relative to other house features and the boiler manufacturer's instructions will also advise.

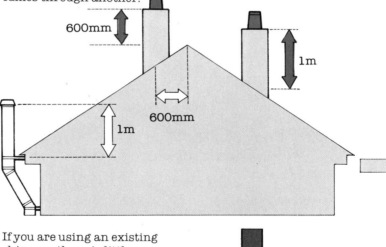

If you are using an existing chimney, there is little you can do to alter it, but recommendations concerning distances above roofs should be followed where prefabricated chimneys pass through or are terminated near roofs. The flue supplier's instructions usually specify minimum distances.

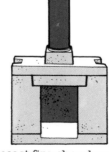

Precast fire-chambers are available for building into houses which lack a fireplace or chimney.

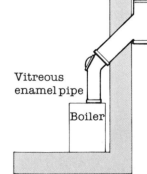

An insulated prefabricated chimney is easy to install following the manufacturer's instructions. They are available for all fuel types.

The regulations

Various rules and regulations may affect the installation of a new central-heating system. Some have the force of law while others are merely recommendations. They should always be followed, as they are for your own safety.

1 Planning permission This covers what can be built, and where. The only situation where it applies to central heating is usually the installation of an oil storage tank, and possibly the siting of an external flue. Contact the local planning officer for advice.

2 Building regulations These cover the detailed requirements of how something is built. The only parts of a central-heating system likely to be affected are the installation of a solid-fuel room-heater and various types of flue. Again, contact your local planning officer for advice.

3 Gas regulations It is unlawful for unauthorized persons to interfere with the gas supply; for this reason gas connections to boilers must be made by an approved gas fitter. Check first with the Gas Board that your own supply pipe is adequate and that the meter is suitable for the boiler you intend to use. If they are not, make arrangements with the Gas Board to have the work done.

You are allowed to make up the gas supply pipe from the boiler to the service pipe, *but not to connect it*; for this, only capillary joints are allowed. Any union joints to the boiler should be of the approved cone type, but as these are usually supplied with the boiler there should be no problems.

4 Water regulations The installation of a central-heating system, if carried out properly, will not usually contravene the local water by-laws, but some authorities don't like pipes buried in floor screeds. A telephone call to the local Water Board is a sensible idea.

5 Electricity Board regulations The usual standard for electrical wiring is the IEE regulations; this is a detailed list of recommendations published by The Institute of Electrical Engineers. If manufacturers' instructions and the details shown in this book are carefully followed, you should have a trouble-free installation.

Upstairs, underfloor wiring should be run through holes drilled in the joists at least 50mm below the top, never over them.

Where wiring is likely to be subject to excessive heat, heat-resisting cable should be used (although this should not be necessary where cables run near

heating pipes). Insulation can be wrapped around pipes where cables unavoidably cross them.

Finally, no cables should be left exposed on walls — fit plastic channelling or conduit for safety; where cables pass through boiler casings or pattress boxes on the back of sockets, rubber grommets should be fitted.

Ventilation for conventional flue boilers

Vent	Area per kW of boiler (mm²)
High level	450
Low level	900

Two permanent air vents, one at low level and one at high level must be fitted in the room where the boiler is sited. Keep the vents away from any pipes to avoid frost damage. Boiler manufacturer's installation instructions will also carry recommendations.

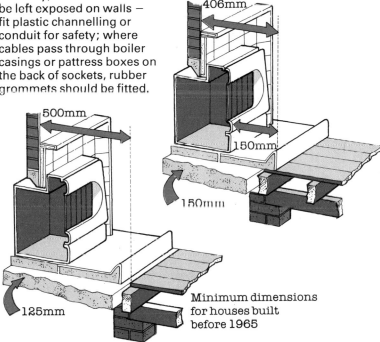

Minimum dimensions for houses built after 1965

406mm

150mm

500mm

150mm

125mm

Minimum dimensions for houses built before 1965

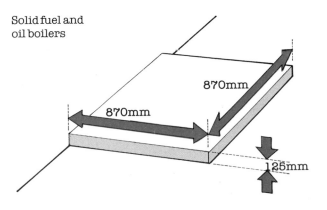

Solid fuel and oil boilers

870mm

870mm

125mm

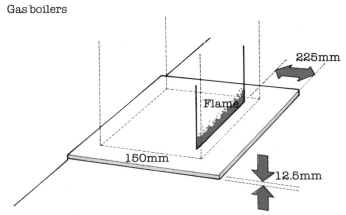

Gas boilers

225mm

Flame

150mm

12.5mm

Minimum non-combustible hearths required for freestanding boilers

Heat-loss calculation charts

These heat-loss charts can be used to calculate the losses for each room in detail. The totals can then be marked on the house heat-loss calculator. Write in pencil in case of errors, or use the charts merely as a guide and write your calculations in a separate notebook. See pages 36-9 for information on how to make the calculations.

	Walls	Length	Height (Width)	Area	Ambient design temperature °C					Total heat loss
					Minus window area	Total	U value	Temp. diff.	Heat loss	
Room	A	×	=	−	=	×	×	=		Add all heat losses together
	B	×	=	−	=	×	×	=		
	C	×	=	−	=	×	×	=		
Required room temp. °C	Ceiling	×	=	−	=	×	×	=		
	Floor	×	=	−	=	×	×	=		
	Window	×	=	−	=	×	×	=		
		×	=	−	=	×	×	=		
Air changes ×	Cubic capacity	×	Air heat loss factor 0.33 ×	Temp. difference			=			
Room	A	×	=	−	=	×	×	=		
	B	×	=	−	=	×	×	=		
	C	×	=	−	=	×	×	=		
Required room temp. °C	Ceiling	×	=	−	=	×	×	=		
	Floor	×	=	−	=	×	×	=		
	Window	×	=	−	=	×	×	=		
		×	=	−	=	×	×	=		
Air changes ×	Cubic capacity	×	Air heat loss factor 0.33 ×	Temp. difference			=			
Room	A	×	=	−	=	×	×	=		
	B	×	=	−	=	×	×	=		
	C	×	=	−	=	×	×	=		
Required room temp. °C	Ceiling	×	=	−	=	×	×	=		
	Floor	×	=	−	=	×	×	=		
	Window	×	=	−	=	×	×	=		
		×	=	−	=	×	×	=		
Air changes ×	Cubic capacity	×	Air heat loss factor 0.33 ×	Temp. difference			=			
Room	A	×	=	−	=	×	×	=		
	B	×	=	−	=	×	×	=		
	C	×	=	−	=	×	×	=		
Required room temp. °C	Ceiling	×	=	−	=	×	×	=		
	Floor	×	=	−	=	×	×	=		
	Window	×	=	−	=	×	×	=		
		×	=	−	=	×	×	=		
Air changes ×	Cubic capacity	×	Air heat loss factor 0.33 ×	Temp. difference			=			
Room	A	×	=	−	=	×	×	=		
	B	×	=	−	=	×	×	=		
	C	×	=	−	=	×	×	=		
Required room temp. °C	Ceiling	×	=	−	=	×	×	=		
	Floor	×	=	−	=	×	×	=		
	Window	×	=	−	=	×	×	=		
		×	=	−	=	×	×	=		

			Ambient design temperature °C							
	Walls	Length	Height (Width)	Area	Minus window area	Total	U value	Temp. diff.	Heat loss	Total heat loss
Room	A	×	=	−	=		×	×	=	Add all heat losses together
	B	×	=	−	=		×	×	=	
	C	×	=	−	=		×	×	=	
Required room temp. °C	Ceiling	×	=	−	=		×	×	=	
	Floor	×	=	−	=		×	×	=	
	Window	×	=	−	=		×	×	=	
		×	=	−	=		×	×	=	
Air changes × Cubic capacity × Air heat loss factor 0.33 × Temp. difference =										
Room	A	×	=	−	=		×	×	=	
	B	×	=	−	=		×	×	=	
	C	×	=	−	=		×	×	=	
Required room temp. °C	Ceiling	×	=	−	=		×	×	=	
	Floor	×	=	−	=		×	×	=	
	Window	×	=	−	=		×	×	=	
		×	=	−	=		×	×	=	
Air changes × Cubic capacity × Air heat loss factor 0.33 × Temp. difference =										
Room	A	×	=	−	=		×	×	=	
	B	×	=	−	=		×	×	=	
	C	×	=	−	=		×	×	=	
Required room temp. °C	Ceiling	×	=	−	=		×	×	=	
	Floor	×	=	−	=		×	×	=	
	Window	×	=	−	=		×	×	=	
		×	=	−	=		×	×	=	
Air changes × Cubic capacity × Air heat loss factor 0.33 × Temp. difference =										
Room	A	×	=	−	=		×	×	=	
	B	×	=	−	=		×	×	=	
	C	×	=	−	=		×	×	=	
Required room temp. °C	Ceiling	×	=	−	=		×	×	=	
	Floor	×	=	−	=		×	×	=	
	Window	×	=	−	=		×	×	=	
		×	=	−	=		×	×	=	
Air changes × Cubic capacity × Air heat loss factor 0.33 × Temp. difference =										
Room	A	×	=	−	=		×	×	=	
	B	×	=	−	=		×	×	=	
	C	×	=	−	=		×	×	=	
Required room temp. °C	Ceiling	×	=	−	=		×	×	=	
	Floor	×	=	−	=		×	×	=	
	Window	×	=	−	=		×	×	=	
		×	=	−	=		×	×	=	

Itemized checklist

Use this checklist along with your isometric drawing to detail the requirements of your new heating system. You are most likely to underestimate on 15mm pipe and 15mm elbows, so add 50 per cent to what you reckon you will need. Items not used can be returned to the supplier for a refund.

		£	p
1 Boiler Make and model Output (kW) Fuel type Flue type			
Flue or prefab. chimney Dia. (mm) Length (m) Flue spigot or vitreous enamel pipe			
Optional pump kit Optional programmer kit			
1a Oil Storage Tank Capacity and size Tank kit (or list separate fittings) Oil line			
2 Hot-water Cylinder Capacity and size Conversion unit Immersion heater (kW)			
3 Feed and Expansion Tank Capacity and size Ball valve and tank connector			
4 Radiators Make and type (single, double, convector or other) Height Length (mm)			
4a Radiator valves Lockshield valves Thermostatic radiator valves			
5 Pump Make and model Head Isolating valves (pipe size mm)			
6 Control valves Zone valves Diverter valve (2-way or 3-way) Mechanical hot-water valve (List type, make, model and pipe size)			
7 Pipe Connectors To boiler size and type Blanking plugs To HWC Air vent (Pumped systems) Primary Secondary To feed and expansion tank Tank connectors Tap connectors PVC or copper overflow – tank connector			

8 Pipe Fittings Couplings   15mm   22mm   28mm
Reducing couplings
Elbows
Equal tees
Reducing tees – Branch reduced
        End reduced
        Branch and end reduced			

Total carried forward			

9 **Pipe** Total lengths (m) in 28mm 22mm 15mm PVC pipe for overflow		
10 **Drains** Boiler Cylinder Radiator (drop feed only) Pipework		
11 **Stopcocks/Gate Valves** Pipe size Check valve Regulating valve		
12 **Electrical Controls** Programmer or Programmer kit Make and model Time switch Make and model Room thermostat Cylinder stat Frost stat		
13 **Electrical Sundries** Junction box amp Connection box No. of terminals Cable 1-5 mm² twin and earth Length (m) Cable clips PVC conduit or channel Length (m) Conduit clips		
14 **Insulation** Pipe Type size quantity HWC size Feed and expansion tank size		
15 **Materials** Gas cylinder – make Radiator screws Solder 500g reel Wall plugs Flux Pipe clips Jointing compound Wire wool PTFE tape Densotape Inhibitor Fire cement Asbestos rope		
16 **Hire of Special Tools**		
Total		
Less refunds on over-ordered or exchanged items		
Less cash for scrap copper etc. sold		
Grand total		

Scrap disposal

The chances are that during the installation of your new central-heating system you may have removed old pipework, a cylinder perhaps, and maybe a boiler.

Lead, copper, brass and cast iron are all valued by the scrap recycling industry. A solid-fuel room-heater in good condition can usually be sold easily, either through the local paper, or by means of a card in a local newsagent's window.

An old Victorian fireplace (or the tiles) will almost certainly be worth something – possibly a great deal – and local antique dealers may be interested.

Nearly all old boilers will contain cast iron but they may be too heavy for you to transport to a scrap-metal dealer. Offer it to a local dealer or contact the local waste disposal department.

If you've got a reasonable quantity of lead or copper pipes, some brass fittings or a copper hot-water cylinder, sort them into their various metal types and look for the main scrap-metal dealer for your area. The dealer will usually be located on a nearby industrial estate.

Keeping records

When your new central-heating system is complete and working correctly, collect all your plans, diagrams, manufacturer's installation leaflets and invoices (for possible future guarantee claims), put them into a large strong envelope or folder and store them.

If any future replacements or additions are needed, these documents will be invaluable, especially the detailed wiring diagrams.

Interested friends may enquire about the costs of a DIY installation, and a complete list of prices will be useful to refer to.

Running costs can be reduced by checking over your insulation and seeing where it can be improved, particularly draughtproofing.

Another possibility is varying the time when the heating and hot water are switched on and off by the controller. Slight adjustments here may show significant improvements over a period of time, as also will resetting the room thermostat to a slightly lower temperature.

Sensible use of your controls will give you a comfortable home without breaking the bank.

Metric conversion charts

Inches/millimetres

in		mm
0.04	1	25.4
0.08	2	50.8
0.12	3	76.2
0.16	4	101.6
0.20	5	127.0
0.24	6	152.4
0.28	7	177.8
0.31	8	203.2
0.35	9	228.6
0.39	10	254.0
0.43	11	279.4
0.47	12	304.8
0.51	13	330.2
0.55	14	355.6
0.59	15	381.0
0.63	16	406.4
0.67	17	431.8
0.71	18	457.2
0.75	19	482.6
0.79	20	508.0
0.83	21	533.4
0.87	22	558.8
0.91	23	584.2
0.94	24	609.6
0.98	25	635.0
1.02	26	660.4
1.06	27	685.8
1.10	28	711.2
1.14	29	736.6
1.18	30	762.0
1.22	31	787.4
1.26	32	812.8
1.30	33	838.2
1.34	34	863.6
1.38	35	889.0
1.42	36	914.4

Feet/metres

ft	in		m
3	3	1	0.30
6	7	2	0.61
9	10	3	0.91
13	1	4	1.22
16	5	5	1.52
19	8	6	1.83
23	0	7	2.13
26	3	8	2.44
29	6	9	2.74
32	10	10	3.05
65	7	20	6.10
98	5	30	9.14
131	3	40	12.19
164	0	50	15.24
196	10	60	18.29
229	8	70	21.23
262	6	80	24.38
295	3	90	27.43
328	1	100	30.48

Square feet/square metres

sq ft		sq m
10.8	1	0.09
21.5	2	0.19
32.3	3	0.28
43.1	4	0.37
53.8	5	0.46
64.6	6	0.56
75.3	7	0.65
86.1	8	0.74
96.9	9	0.84
107.6	10	0.93
215.3	20	1.86
322.9	30	2.79
430.6	40	3.72
538.2	50	4.65

Gallons/litres

gal		lit
0.2	1	4.5
0.4	2	9.1
0.7	3	13.6
0.9	4	18.2
1.1	5	22.7
1.3	6	27.3
1.5	7	31.8
1.8	8	36.4
2.0	9	40.9
2.2	10	45.5
4.4	20	90.9
6.6	30	136.4
8.8	40	181.8
11.0	50	227.3
13.2	60	272.8
15.4	70	318.2
17.6	80	363.7
19.8	90	409.1
22.0	100	454.6

Pounds/kilograms

lb		kg
2.2	1	0.5
4.4	2	0.9
6.6	3	1.4
8.8	4	1.8
11.0	5	2.3
13.2	6	2.7
15.4	7	3.2
17.6	8	3.6
19.8	9	4.1
22.0	10	4.5
44.1	20	9.1
66.1	30	13.6
88.2	40	18.1
110.2	50	22.7
132.3	60	27.2
154.3	70	31.8
176.4	80	36.3
198.4	90	40.8
220.5	100	45.4

Temperature

°F	212	175	140	105	95	85	75	70	60	50	40	32
°C	100	80	60	40	35	30	25	20	15	10	5	0

Glossary

A

Air lock Air trapped in pipes or radiators obstructing the flow of water.

Air changes Rate at which air is changed in a room.

Accelerated thermostat/ heat anticipator Hot-wire circuit in a room thermostat to counteract temperature fluctuations.

Ambient temperature Surrounding air temperature. Usually taken as outside temperature when related to heat-loss calculations.

Air eliminator Air valve.

Actuator Non-electric domestic hot-water temperature controller, or the motor part of a motorized valve.

Aquastat Cylinder thermostat.

Agrément certificate Awarded to certain worthy building materials by an independent body.

B

Bypass, bypass valve Circuit incorporated into a system to allow continuous water flow through the boiler to avoid overheating.

Back boiler Cast-iron water container directly heated by a fire.

Building regulations Mandatory conditions governing the alterations to the structure or fabric of a house.

Balancing valve Simple screw-down valve used to reduce the flow through a circuit.

Background heating Low level of heating requiring (at times) an additional source of radiant heat.

Bar, milliBar Units of measurement of pressure.

Ballcock Ball valve such as that fitted in a cistern.

Bleeding Venting trapped air in a radiator or pipe.

Boiler cycling Boiler switching on and off intermittently.

Boss Protruding opening from a boiler or cylinder.

B (continued)

Branch A pipe taken at right angles from the main flow.

British Standard (BS) Usually followed by a number denoting that the item concerned meets the minimum requirements for that standard.

British Standard Pipe (BSP) Measurement of thread size on a boiler or cylinder. Still in use despite metrication.

British Thermal Unit (BTU) Old measurement of energy, now replaced by the kiloJoule and kiloWatt.

Burr Sharp edges on a pipe end caused by cutting.

C

Calorifier The primary heating element in a hot-water cylinder.

Capillary Solder fitting using the principle of capillary attraction.

Catchpit A construction around an oil storage tank to contain the contents in the event of a leak.

Chase Channel cut in plaster in which electric cables are concealed.

Check valve A simple weighted valve fitted in pipework to restrict gravity flow when the pump is off.

Circulator Pump.

Cistern An open-topped water container, usually containing a ballcock. Strictly speaking, all so-called 'tanks' are cisterns.

Class I appliance Solid-fuel and oil-fired boilers.

Class II appliance Gas-fired boilers.

Commissioning The boiler is 'commissioned' when the first lighting takes place after the necessary settings and adjustments have been made.

Compression fitting A fitting where the joint is made by tightening down nuts on the pipe.

Constructional hearth The hearth that is laid when a house is built. It may have a decorative, superimposed (tiled) hearth on top.

Consumer unit The unit connected to the electricity

C (continued)

supply; it contains the household's fuses and main switch.

Crawling board Roof ladder.

Cylinder jacket Thick, insulating cover for the hot-water cylinder.

D

Damp-proof membrane A strip of special felt inserted between the supporting brick piers and an oil storage tank.

Dead leg Long run of pipe to a single hot tap.

Direct cylinder A cylinder in which the directly heated water is drawn off for use.

Double-pole isolator A switch which cuts off the current at both the live and neutral terminals.

Down draught Air flow down a flue instead of up.

Drain cock Drain valve.

Draw-off valve As above.

Duct A large tube, usually rectangular, used for balanced-flue outlets.

E

Elbow Small-radius right-angle bend fitting.

Electrolytic action Corrosion in galvanized tanks caused by copper pipes.

Emitter Technical name for all kinds of radiators.

End-feed fitting Solder fitting which does not contain a ring of solder.

Expansion pipe See feed and expansion pipe.

F

Fascia board The vertical board at the eaves of a house.

Feed and expansion pipe The pipe that supplies cold water to the boiler.

Feed and expansion tank The tank that holds a supply of cold water to replenish the boiler.

Female iron A tapping on a boiler, cylinder, or fitting, with an internal thread. Not necessarily iron.

Fireback Shaped fire-clay

F (continued)

back fitted around most open fires. May be replaced by a cast-iron water jacket in a boiler fire.

Firebox Part of a solid-fuel boiler that contains the burning fuel.

Fire valve A special valve fitted in an oil installation which cuts off the supply of oil in the event of a fire.

Float valve Ball valve.

Flue Construction for venting waste gases to the outside atmosphere.

Flue-liner Special lining inserted in a chimney.

Flue spigot Recessed joint where the flue connects to the boiler.

Forced convection Fan-assisted convector radiator.

Free-standing room-heater A boiler not set into a fireplace.

G

Gas cock A valve which turns through 90 degrees to shut off the gas supply.

Gate valve A screw-down valve containing a metal plate to shut off the water supply.

Gravity circulation Not assisted by the pump, a gravity system relies on the different densities between hot and cold water for circulation.

Gully Outside drain to a sewer.

H

Head, head loss The height of the water in the system or a method of determining the pump required for a particular system.

Heat anticipator See Accelerated thermostat.

Heat-exchanger Part of a boiler where heat from combustion is transferred to the water.

Heat leak The cylinder is usually the heat leak on a gravity circuit, its purpose being to absorb surplus heat without waste, but an unvalved radiator or towel-rail may be used as a substitute.

Heat loss A calculation for determining the heat lost through the structure of the house.

Hexagon wrench Six-sided bent rod used for some types of radiator valve.

High-pressure zone An area around a flue or chimney with higher buildings or trees, preventing its proper operation.

High thermal-capacity boiler A boiler with a heavy cast-iron heat-exchanger.

Hole saw Various types of cutters (power-drill attachment) for making large diameter holes.

Humidifier A water-filled container for replacing moisture in the air removed by central heating. May be electrically operated.

I

ID Internal diameter.

Impeller Pump, or part of a pump.

Index circuit The circuit in a central-heating system creating the greatest resistance to the action of the pump.

Indirect cylinder A hot-water cylinder in which the domestic hot water is separated from, and heated indirectly by, the water in the boiler.

Inhibitor Chemical additive for a central-heating system, formulated to give protection against corrosion and sludge.

Injector tee A special tee designed to avoid back-flow in a system.

Integral ring fitting A fitting with solder rings embedded in the ends.

Intermittent switching Boiler controlled by remote electrical thermostats.

Isolating valve Any valve which shuts off part of a system.

J

Joist Timbers of a floor that support the floorboards.

Joule, kiloJoule Unit of energy.

K

Kelvin Unit of temperature used only by scientists. The centigrade (or celsius) measurement is used for heating.

Kettling Rattling noises from lightweight heat-exchangers in low-water-content boilers.

L

Lagging Insulation on pipes, cylinders and tanks.

Liner See Flue-liner.

Live Usually refers to something carrying electrical current.

Live radiator See Heat leak.

Low thermal capacity Boiler with lightweight heat-exchanger.

Low-water-content Small, lightweight boiler designed to function with a small amount of water.

Lockshield valve Radiator valve, usually with a cover, used to balance the radiator.

LPG Liquid Petroleum Gas.

M

Main (electric) Usually refers to the ring main.

Main (water) Usually refers to the rising main cold-water supply pipe.

Main stopcock The valve which shuts off the whole cold-water supply of a house.

Mains tester Special electrical screwdriver containing a device for testing if current is present at a terminal.

Male iron A tapping on a boiler, cylinder, or fitting, with an external thread. Not necessarily iron.

Manifold The part of a microbore system from which the microbore pipes radiate to the radiators.

Manipulative fitting A type of fitting, not much used in domestic heating, in which the pipes are belled out with a special tool to make a joint.

Mean temperature Temperature halfway between two differing temperatures on, for example, the flow and return pipes to a boiler.

Microbore, minibore System using smaller pipes than the normal small bore minimum of 15mm.

Monobloc Made in one piece.

Motorhead The electrical motor part of a motorized valve.

Motorized valve A control valve which is open and shut by an electric motor.

Multifuel boiler A boiler which can be fired by more than one type of fuel.

N

Negative pressure The system is under negative pressure when the pump is sited in the return and not the flow.

Newton Unit of force measurement.

Nipple A connector, threaded at both ends for joining two similar connectors.

Night set-back A modification to a room thermostat for running heating at a low rate during the night instead of completely shutting it off.

O

Obtuse tee Specially shaped tee to avoid back flow through the branch.

OD Outside Diameter.

Olive Sealing ring on a compression fitting.

Open vent Open pipe rising from the boiler and terminating above the feed and expansion tank.

OSV Open Safety Vent.

P

Pascal Unit of force.

Pipe cutter Special tool for accurate pipe cutting.

Pipe sizing Using tables to determine the diameter of pipe needed to carry a specific heat load.

Pipe thermometer Special thermometer used to denote the temperature of water flowing through a pipe for system adjustment.

Pipe wrench Gripping tool for turning pipe.

Planning permission Approval that may be required for proposed work, according to local by-laws.

Positive pressure The system is under positive pressure when the pump is sited in the flow pipe.

Pressurized system A central-heating system completely sealed to the atmosphere. Not covered in this book due to the special equipment required.

Pressure jet Type of oil burner.

Primary The pipework that connects the heating water to the cylinder.

Priority system A system employing a motorized diverter valve which will supply either the heating or hot-water circuits, but not both. Priority is usually given to the hot-water circuit.

Programmer Sophisticated clock which governs the action of all the components in a heating system.

PTFE tape Very thin, white, plastic tape used for sealing threaded joints.

Pump performance graph A graph supplied with the pump showing the flow through the pump at different settings.

Pump pipe A short piece of pipe used in place of the pump when flushing the system.

Pump sizing Selecting the pump from calculations made from a particular system.

R

Radiant Heat Direct heat as opposed to convected heat.

Radiator sizing Determining the size of a radiator from heat loss calculations.

Range rated Covering a range of outputs from one boiler. The boiler firing rate is adjusted at the time of commissioning to suit the

particular system.
Remote sensor Separate temperature sensing device for hot-water cylinders or thermostatic radiator valves.
Ring main Method of supplying a number of socket outlets throughout a house from a ring of cable.
Rising main The main cold-water supply into the house.

S

Screed The finished concrete surface on a solid floor.
Secondary pipework The pipework that carries the hot water used in the house to and from the cylinder.
Sensor See Remote sensor.
Sight gauge An oil tank accessory that shows the tank's contents at a glance.
Sludge valve Manual valve fitted to an oil tank to drain off sludge.
Small bore heating system The standard system using a minimum pipe diameter of 15mm.

Smoke-control zone

If you live in a smoke-control area, you can only burn smokeless fuel. If the area in which you live is being made a smokeless zone, and you have to change your boiler, you may be able to claim a grant to cover most of the cost of replacement. Enquire at the local authority offices.

Soffit board The horizontal board at the eaves of a house.

Solar gains Heat gained by the sun shining through windows.
Solid fuel Generally taken to mean coal and processed coal products but can include wood and peat.
Solvent-welded joints Used in plastic plumbing.
Space heaters A fire that heats the room in which it is situated. Room-heaters with back boilers often quote two figures, one for space heating, the other for radiators and hot water.
Spigot See Flue spigot.
Static head The height of water in the system when the pump is turned off.
Stillson wrench See Pipe wrench.
Stopcock Manual shut-down valve.
Suspended floor Floorboards on joists.
Swarf Metal filings.
Syphon The flow of water created by suction.
Service pipe Main water supply to the house.
Skirting heating Long, low heating units replacing skirting boards.
Smoke eater A solid-fuel boiler with a secondary combustion chamber for extra efficiency.
Stop end A fitting which blanks off a pipe end.

T

Tail Part of a radiator valve that screws into the radiator.
Tank connector Fitting designed to pass through the side of a polythene tank.
Tap connector Fitting that connects a pipe to a tank connector.
Tariff Table of fuel costs.

Tee Connector joining two pipes at right angles to each other.
Temperature drop The difference in temperature between flow and return pipes.
Terminal Wiring block for connecting an electrical item.
Thermostatic radiator valve A special valve which reacts to air temperature and as a result reduces the flow of hot water through a radiator.
Time switch Mechanical clock for switching a pump on and off.

U

U values Tables of thermal transmittance.
Union wrench Special tool for fitting some kinds of radiator valves.

V

Velocity of water The speed at which water can be pumped through small bore pipes.
Vent Releases air to the atmosphere.
Vermiculite concrete A lightweight insulating concrete.
Vitreous enamel pipe Decoratively finished flue-pipe for use where a boiler flue is exposed in a room.

W

Warning pipe Overflow pipe.
Wet system Descriptive term covering the common water-boiler and radiator system as opposed to a

warm-air system.
Wire strippers Special tool for quickly removing the insulation sheaths from electrical wiring.

Y

Yorkshire fitting Integral solder ring fitting.

Z

Zone valve A motorized valve used to control the heating of various zones (ground floor and first floor). A more common application, though, is one zone valve controlling heating and another controlling hot water.

Abbreviations used by various organizations

CAS   Coal Advisory Service (Northern Ireland)
CORGI   Confederation of Registered Gas Installers
CUC   Coal Utilization Council
DOBETA   Domestic Oil Burning Equipment Testing Association
DSFAAS   Domestic Solid Fuel Appliances Approval Scheme
HVCA   Heating and Ventilating Contractors Association
IEE   Institute of Electrical Engineers
CIBS   Chartered Institute of Building Services
NCB   National Coal Board
NFC   National Fireplace Council
NWC   National Water Council
SFAS   Solid Fuel Advisory Service

Acknowledgments
Aga-Rayburn: pages 14, 16, 17, 20, 21 (below); Conex-Sambra Ltd: pages 2, 8, 26; Drayton Controls (Engineering) Ltd: pages 4-5, 30 (top left), 31 (right); Dunsley Heating Appliance Co. Ltd: page 17 (below right); Fin-Rad Ltd: page 23 (below right and left); Grundfos Pumps Ltd: page 28 (above); Honeywell Control Systems Ltd: pages 30 (top right, centre and below), 31 (left); Kedddy Home Improvements Ltd: page 19 (below);

Mysons plc: pages 15, 22 (below right), 23 (above and centre); Potterton International Ltd: pages 12 (below, left and right), 13 (above right), 18 (above); Stelrad Group Ltd: pages 12 (above), 13 (below), 18 (below), 22 (above and below left); Thorn EMI Heating Ltd: page 28 (below); TI Parkray Ltd: page 17 (below left); Trianco Redfyre Ltd: page 19 (above); UA Engineering Ltd: page 21 (above); The Wednesbury Tube Co.: pages 25, 32, 33.

Index

Figures in italics refer to illustrations

AGA cooker, 20, *21*
air locks, clearing, 83, 95
air valve, 43, 89, *89*

balanced-flue boiler, 7, 12, *10*, 13
balancing, system, 95
balancing valve, 89
boiler
  choice of, 7, 13, 39
  commissioning, 94
  control systems, 60
  installing, 72-3, 74-5, *72, 73, 74, 75*
  replacing, 97
  siting, 7, 43
  see also under specific types
boiler allowance, 39
boiler thermostat, 30
building regulations, 103
bypass valve, 89, *89*

capillary fittings, 24, *24, 25*
  installing, 70, *70*
checklist, itemized, 106-7
check valve, *31, 89*
chimney, 7, 102, *102*
commissioning, 94-5
compression fittings, 26-7, *26, 27*
  installing, 71, *71*
control layouts, 30-1, 60, *30, 31, 60, 61*
control valve, 88-9, *88, 89*
controller, 10, 31, *11, 30*
convector radiator, 23, 96, *96*
conversion unit, 93
cooker/boiler, 20-21, *20, 21*
corrosion inhibitor, 22, 98
cylinder circuit, 10
cylinder thermostat, 10, 30, *11*

design, system, 36-9
diverter valve, 88, *88*
domestic hot-water requirement, 39
double radiator, 22
drain valve, 43, 89, *89*
draught-proofing, 101, *101*

electrical connections, 90-91, *90, 91*
electrical fault-finding, 98
Electricity Board regulations, 103

fault-finding, 98
feed and expansion pipes, 43

feed and expansion tank, 84-5, *10, 84, 85*
fire precautions, 81
fireplace surround, removing, 73
floorboards, removing, 78-9, *78, 79*
floor-standing boiler, 12
flow rates, 47, 49
flue-liner, 76, *76*
flues, 15, 102, *102*
fuel, choice of, 6-7

gas boiler, 6, 7, 12-13, *12, 13*
  installing, 74-5, *74, 75*
gas regulations, 103
gas room-heater, 12, 18, *18*
gas supply connections, 75
glossary, 109-11
gravity primary pipes, 20, 43
gravity system design, 7, 52

head loss, 49
heat loss calculations, 36-8
  charts, 104, 105
heat requirement, total, 38
heating circuits, 45
heating system operation, 10
high head pump, 29, *29*
hot-water control valves, 88, *88*
hot-water cylinder, 82, *11*
  replacing, *83*
hot-water system operation, 10
house structures, 36

IEE Regulations, 103
immersion heater, 83
index circuit
  microbore, 56
  smallbore, 49-51
indirect cylinder, 93
insulation, 101, *101*
isometric drawings, 46

layouts, alternative design, *62, 63*
leaks, curing, 98

manifold, microbore, 32, 92, *32*
measuring the house, 41
metric conversion charts, 108
microbore/minibore system, 32-3
  design, 54-7
  installing, 92-3, *92, 93*
  manifold, 32, 92, *32*
minimum flow requirements, 95

oil boiler, 6-7, 15, *15*

oil storage tank, 77, *77*
open fires, 17, *17*
open vent, 43, *10*

panel convector, 22
pipes
  bending, 67, *67*
  installing under floors, 79-81, *79, 80, 81*
  layouts, 43, 45
  sizing, 47-8
pipework, microbore, 32, *32*
planning permission, 103
planning the work, 94-5
plans, drawing, 41-5
pressure-jet boiler, 15
propane gas boiler, 7
pump, 28-9, *28, 29*
  designing for, 60
  fitting, 86-7, *86, 87*
  for microbore systems, 56
  replacing, 99, *99*
pump performance calculations, 49, 51
  graph, *28*

radiator valves
  fitting, 69
  microbore, 93
  types of, *69*
radiators, 22-3, *11, 22, 23*
  checking, 98
  convector, 23, 96, *23, 96*
  difficult areas, 96
  fitting, 68-9, *68, 69*
  painting, 97, *97*
  removing, 97, *97*
  replacing, 98
  siting, 43
  sizing, 40
record keeping, 108
regulations, 103
room-heater, 16-17, 18, *16, 17, 18*
room thermostat, 10, 30, *11, 30, 31*

safety precautions
  blowlamps, 81
  electrical, 91
  roof-working, 76
  when installing, 75
safety valve, 89, *10, 89*
scrap disposal, 108
sealed system, 62
sensor valve, remote, 88
servicing and maintenance, gas-boiler, 13
skirting radiator, 23, *23*
smoke control zone, 20, 111
solder-ring fittings, 24
solid-fuel boiler, 6, 7, 14, *14*
solid-fuel room-heater, 16-17, *16, 17*
  installing, 72-3, *72, 73*

solid-fuel system design, 52
storage tank, replacing, 100, *100*
system balancing, 95
system requirements checklist, 94-5

temperature drop, 47
temperature, water, 47
threaded fittings, 71
tools, 66-7, *66, 67*

valves, 30-31, *31*
  air, 43, 89, *89*
  balancing, 89
  bypass, 89, *89*
  check, *31, 89*
  control, 88-9. *88, 89*
  diverter, 88, *89*
  drain, 43, 88, *89*
  microbore, 33, 54 *33*
  motorized, *30*
  thermostatic, *31*
ventilation requirements, 14
vermiculite concrete, 72

wallflame boiler, 15
wall-hung boiler, 12, 13, 74-5, *13, 74, 75*
water regulations, 103
waterway thermostats, 16
wiring, electrical, 90-91, *90, 91*
wiring diagrams, 58, *59*
wood-fired boiler, 7, 19, *19*

zone valve, 10, 88, *11, 88*